American Literature Readings in the 21st Century

Series Editor: Linda Wagner-Martin

American Literature Readings in the 21st Century publishes works by contemporary critics that help shape critical opinion regarding literature of the nineteenth and twentieth century in the United States.

Published by Palgrave Macmillan:

Freak Shows in Modern American Imagination: Constructing the Damaged Body from Willa Cather to Truman Capote
By Thomas Fahy

Women and Race in Contemporary U.S. Writing: From Faulkner to Morrison
By Kelly Lynch Reames

American Political Poetry in the 21st Century
By Michael Dowdy

Science and Technology in the Age of Hawthorne, Melville, Twain, and James: Thinking and Writing Electricity
By Sam Halliday

F. Scott Fitzgerald's Racial Angles and the Business of Literary Greatness
By Michael Nowlin

Sex, Race, and Family in Contemporary American Short Stories
By Melissa Bostrom

Democracy in Contemporary U.S. Women's Poetry
By Nicky Marsh

James Merrill and W.H. Auden: Homosexuality and Poetic Influence
By Piotr K. Gwiazda

Contemporary U.S. Latino/a Literary Criticism
Edited by Lyn Di Iorio Sandín and Richard Perez

The Hero in Contemporary American Fiction: The Works of Saul Bellow and Don DeLillo
By Stephanie S. Halldorson

Race and Identity in Hemingway's Fiction
By Amy L. Strong

Edith Wharton and the Conversations of Literary Modernism
By Jennifer Haytock

The Anti-Hero in the American Novel: From Joseph Heller to Kurt Vonnegut
By David Simmons

Indians, Environment, and Identity on the Borders of American Literature: From Faulkner and Morrison to Walker and Silko
By Lindsey Claire Smith

The American Landscape in the Poetry of Frost, Bishop, and Ashbery: The House Abandoned
By Marit J. MacArthur

Narrating Class in American Fiction
By William Dow

The Culture of Soft Work: Labor, Gender, and Race in Postmodern American Narrative
By Heather J. Hicks

Cormac McCarthy: American Canticles
By Kenneth Lincoln

Elizabeth Spencer's Complicated Cartographies: Reimagining Home, the South, and Southern Literary Production
By Catherine Seltzer

New Critical Essays on Kurt Vonnegut
Edited by David Simmons

Feminist Readings of Edith Wharton: From Silence to Speech
By Dianne L. Chambers

The Emergence of the American Frontier Hero 1682–1826: Gender, Action, and Emotion
By Denise Mary MacNeil

Norman Mailer's Later Fictions: Ancient Evenings through Castle in the Forest
Edited by John Whalen-Bridge

Fetishism and its Discontents in Post-1960 American Fiction
By Christopher Kocela

Language, Gender, and Community in Late Twentieth-Century Fiction: American Voices and American Identities
By Mary Jane Hurst

Repression and Realism in Postwar American Literature
By Erin Mercer

Writing Celebrity: Stein, Fitzgerald, and the Modern(ist) Art of Self-Fashioning
By Timothy W. Galow

Bret Easton Ellis: Underwriting the Contemporary
By Georgina Colby

Amnesia and Redress in Contemporary American Fiction: Counterhistory
By Marni Gauthier

Vigilante Women in Contemporary American Fiction
By Alison Graham-Bertolini

Queer Commodities: Contemporary US Fiction, Consumer Capitalism, and Gay and Lesbian Subcultures
By Guy Davidson

Reading Vietnam Amid the War on Terror
By Ty Hawkins

American Authorship and Autobiographical Narrative: Mailer, Wildeman, Eggers
By Jonathan D'Amore

Readings of Trauma, Madness, and the Body
By Sarah Wood Anderson

Intuitions in Literature, Technology, and Politics: Parabilities
By Alan Ramón Clinton

African American Gothic: Screams from Shadowed Places
By Maisha Wester

Exploring the Limits of the Human through Science Fiction
By Gerald Alva Miller Jr.

A Companion to David Foster Wallace Studies
Edited by Marshall Boswell and Stephen J. Burn

The Middle Class in the Great Depression: Popular Women's Novels of the 1930s
By Jennifer Haytock

Charles Bukowski, King of the Underground: From Obscurity to Literary Icon
By Abel Debritto

Urban Space and Late Twentieth-Century New York Literature: Reformed Geographies
By Catalina Neculai

Revision as Resistance in Twentieth-Century American Drama
By Meredith M. Malburne-Wade

Rooting Memory, Rooting Place: Regionalism in the Twenty-First-Century American South
By Christopher Lloyd

Kate Chopin in Context: New Approaches
Edited by Heather Ostman and Kate O'Donoghue

The Non-National in Contemporary American Literature: Ethnic Women Writers and Problematic Belongings
By Dalia M. A. Gomaa

The Non-National in Contemporary American Literature

Ethnic Women Writers and Problematic Belongings

Dalia M. A. Gomaa

palgrave
macmillan

THE NON-NATIONAL IN CONTEMPORARY AMERICAN LITERATURE

First published 2016 by
PALGRAVE MACMILLAN

Palgrave Macmillan in the UK is an imprint of Macmillan Publishers Limited, registered in England, company number 785998, of Houndmills, Basingstoke, Hampshire, RG21 6XS.

Palgrave Macmillan in the US is a division of Nature America, Inc., One New York Plaza, Suite 4500, New York, NY 10004-1562.

Palgrave Macmillan is the global academic imprint of the above companies and has companies and representatives throughout the world.

Hardback ISBN: 978–1–137–50286–5
E-PUB ISBN: 978–1–137–49625–6
E-PDF ISBN: 978–1–137–49626–3
DOI: 10.1057/9781137496263

Distribution in the UK, Europe and the rest of the world is by Palgrave Macmillan®, a division of Macmillan Publishers Limited, registered in England, company number 785998, of Houndmills, Basingstoke, Hampshire RG21 6XS.

Library of Congress Cataloging-in-Publication Data

Names: Gomaa, Dalia M. A., 1973–
Title: The non-national in contemporary American literature : ethnic women writers and problematic belongings / Dalia M.A. Gomaa.
Description: New York : Palgrave Macmillan, 2016. | Series: American literature readings in the twenty-first century | Includes bibliographical references and index.
Identifiers: LCCN 2015022794 | ISBN 9781137502865 (hardback)
Subjects: LCSH: American literature—Minority authors—History and criticism. | American literature—Women authors—History and criticism. | American literature—20th century—History and criticism. | American literature—21st century—History and criticism. | National characteristics, American, in literature. | Belonging (Social psychology) in literature. | BISAC: LITERARY CRITICISM / General. | LITERARY CRITICISM / Feminist. | LITERARY CRITICISM / American / General. | LITERARY CRITICISM / Women Authors.
Classification: LCC PS153.M56 G66 2016 | DDC 810.9/8—dc23 LC record available at http://lccn.loc.gov/2015022794

A catalogue record for the book is available from the British Library.

To the memory of my mother
and
to my father and brother

Contents

Acknowledgments

I am deeply grateful to many colleagues, friends, and family members whose guidance and support have made this book possible. I have been fortunate to work closely with Kumkum Sangari. Her insights and compassion have opened up for me new ways of thinking and perception. I am also thankful to Gwynne Kennedy for her generosity and gracefulness in all and every occasion I asked for her help or advice. My thanks, as well, go to Kristie Hamilton, who told me very early on: "one day, your book will be on my bookshelf." Her optimism and sense of humor lightened up many tough times I encountered. Thanks to this intellectual trinity, I have grown up both academically and personally.

I am also indebted to the vibrant and diverse communities of scholars and friends I have worked with in both the English Department and the Women's and Gender Studies Department in the University of Wisconsin, Milwaukee. Special thanks go to Brenda Cardenas, Xin Huang (my writing pal), Drew Anastasia, Angy Singh, Eric Herhuth, and Kate Olson Nesheim.

I owe Frances Aparicio at Northwestern University, Juanita Del Toro, and Sam Mitrani a huge thank you. Frances gave me the opportunity to publish my first scholarly essay about Latinos/as in the Arab world in The Routledge Companion to Latino/a Literature. Juanita and Sam's friendship is beyond my words. I also want to thank Nouri Gana for his suggestions, and Pauline Kaldas for offering me the chance to interview her. My thanks also go to Susan Friedman, Evelyn Alsultany, Maria Herrera-Sobek, and Amritjit Singh for their insightful remarks about the manuscript. I am thankful to my friends and colleagues in the English Department in the University of Cairo for their companionship.

I would like to extend my thanks to the Palgrave Macmillan team, Brigitte Shull, who believed in the manuscript, and to Ryan Jenkins, who worked with me patiently and tirelessly throughout the publication process.

Thanks also to the anonymous reviewer whose valuable suggestions sharpened the theoretical framework of my study, and to the copyeditor, who meticulously read the manuscript. My thanks go to the *South Asian Review Journal* and the anonymous editor who gave me the opportunity to publish my essay on Bapsi Sidhwa.

I dedicate this book to my mother, whose presence, as well as her memory, continues to inspire me. Our unique friendship has been my goldmine for prudence, kindness, and humor. Thanks also to my father who has supported and encouraged my pursuit of knowledge and education in every possible way. I am thankful to my brother, Mohamed, for looking after me. I am grateful to my family members for their continued care and support.

Introduction

> *The transnational turn in American Studies has been crucial in decentering the tenacious model of the nation as the basic unit of knowledge production. It traces alternative spaces and modes of belonging to collectivities not subsumed by the nation-state... and it reconceives immigration as multidirectional movements.*
>
> —*Amy Kaplan, 11–12*

In the presidential address to the American Studies Association in 2003, Amy Kaplan underlined the necessity of reconceptualizing national identity in American Studies in light of a newer understanding of the United States and Americanness as relational concepts, constructed *nationally* within the geographical boundaries of the United States and *trans*nationally outside of that space. This creates the possibility of the formation of problematic, sometimes paradoxical affiliations and identifications, which I call non-national, and through its lens I examine narratives by Arab-American, Chicana, Indian-American, Pakistani-American, and Cuban-American women writers. This study problematizes essential concepts that are imminent to the formation of the nation: national consciousness, national time, national space, and national belonging in texts by Diana Abu-Jaber (*The Language of Baklava*, 2005), Laila Halaby (*West of the Jordan*, 2003), Pauline Kaldas (*The Time between Places: Stories that Weave In and Out of Egypt and America*, 2010), Alia Yunis (*The Night Counter*, 2009), Bapsi Sidhwa (*An American Brat*, 2006), Cherríe Moraga (*The Last Generation*, 1993), Jhumpa Lahiri (*The Namesake*, 2003), and Cristina Garcia (*The Agüero Sisters*, 1997). I work my analysis of the non-national through two theoretical frameworks that are interrelated: the transnational approach in American Studies and Arab-American Studies.

The Non-national in American Studies

The nation and the national have been central concepts in American Studies and they have been recently problematized and critiqued by new Americanist scholars.[1] New Americanist scholars such as Donald Pease, Amy Kaplan, Annette Kolodny, Myra Jehlen, Janice Radway, and John Carlos Rowe, to name only a few, critique national narratives that developed out of the consensus politics of the post–Cold War era that universalized notions of the nation-state and national subject. As counter narratives to national narratives, what Pease in *National Identities* (1992) calls "postnational narratives" contest these social arrangements that "produced national identities by way of a social symbolic order that systematically separated an abstract, disembodied subject from resistant materialities, such as race, class, and gender" (3).

Postnational narratives rewrite the genealogy of national identity in American Studies to locate it within social movements and historical events. By way of doing this, new Americanists rehistoricize literary texts written during the decades of "literary nationalism"[2]—the mid-nineteenth century—that emphasized notions of manifest destiny, expansion, exceptional, and isolation. New Americanists' rereadings of texts from these decades restore the correlation between the political context and literary texts in nineteenth-century America. For example, John Carlos Rowe in his rereading of Herman Melville's *Typee*—in "Postnationalism, Globalism, and the New American Studies" (2002)—explores "the relationship between domestic policies of southern slaveholding and the extraterritorial [outside the North American continent] policies of U.S. colonization that began as early as the war of 1812" (256). Making this connection between the US domestic and foreign policies situates US history of slavery within its colonial past. In a different sense, Amy Kaplan and Donald Pease in *Cultures of United States Imperialism* link the domestic and the foreign to foreground "the multiple histories of continental and overseas expansions, conquest, conflict, and resistance which have shaped the cultures of the United States, and the cultures of those it has dominated within and beyond its geopolitical boundaries" (4). Along the same lines, in what they describe as "hemispheric" American Studies, Caroline Levander and Robert Levine take a "comparativist approach to consider the overarching shape and texture of American literary and cultural history." Hemispheric Studies moves beyond the nation to "consider regions, areas, and diasporan affiliations that exist apart from or in conflicted relation to the nation" (2).

Within these studies, the term "non-national" has been used in different ways. Levander and Levine have used it to criticize how narratives by African Americans, Native Americans, Asian Americans, and Latinos/Chicanos have been excluded from American Studies because they do not fit in a nation-based teleological schema (1). Radway in "What's in a Name" (2002) uses the term "non-national" as part of her argument to change the name of the field from American Studies to either "United States Studies" or "Society for Intercultural Studies" (60, 64). She holds that these alternative names expand a narrow definition of America/Americanness beyond the geographical boundaries of the United States and contest a coherent nationalism. By the same token, Russ Castronovo uses the term to argue for a different pedagogy in American Studies programs and to critique the current pedagogy that emphasizes US-centrism in American Studies. Non-national pedagogy, as Castronovo holds, encourages an understanding of the United States within a wider context than the narrow understanding of the United States as a nation-state. In two scholarly studies, Lori McNulty—in *Contagious Poetics: Rumor, Ritual and Resistance in Zora Neale Hurston's "Tell My Horse"* (1999)—and Chandan Reddy—in *The Migrant Present: Alienage, Race and the Politics of Black Internationalism, 1898–1953* (2004)—use "non-national" to describe, respectively, color-based affiliations between blacks in Haiti and Saint Domingue and blacks in the African continent, and the racialization of African American and Chinese-American laborers. Lori McNulty examines voudoun communities as sites of resistance in colonial Haiti and Saint Domingue and argues that it is a community-based movement rather than a nation-based one. It is a form of collectivity that is outside and before the Eurocentric tradition of forming collectivities/communities predicated on a notion of the nation. It is rather a ritual of resistance among displaced West African slaves. Chandan Reddy discusses the material conditions of black and Chinese laborers at the time of the United States expanding its imperial policies in Asia and the Caribbean—in the mid-nineteenth century—that rendered blacks and Asians the "outsiders within" (10). In that sense, the "non-national" is "a terrain of intersecting racial histories of colonialism, imperialism, displacement and migration" (16). Still in the Americas, but outside the United States, Patricia Eugenia Varas, in her study, uses the term non-national to describe literature in Ecuador that has its roots in Europe. Varas concludes that Ecuadorian literature that is rooted in Europe is non-national. On a wider scale, Pascal Casanova in *The World Republic of Letters* (2007) uses "non-national" to describe literature written outside its country of origin (e.g., Gertrude Stein writing about the United States while in France) or literature claimed to be universal, appealing to all other cultures (e.g., William Shakespeare).

In this book, I examine the perception of America and Americanness from inside the United States as well as from outside it. The incidents in the literary texts I include in my study take place in the United States and also refer to or take place in precolonial Mexico, Cuba, Jordan, the West Bank, India, and Pakistan. They reveal the intertwined political and cultural histories of the United States, Central America, the Arab region, and South Asia. Despite their different histories in the United States, the groups of writers and texts in here share the themes of migration and problematic belongings. I deliberately chose to read these texts together because they are about mobility and migrations to the United States in the late twentieth century in relation to women of color. I analyze these texts through the vantage point of the non-national, which, I argue, lies at the intersection of the national and the transnational. Each of the minority groups that I include in my study has been excluded or subordinated either as "wet backs" and illegal immigrants (Mexicans/Latinos (as)), as potential terrorists and aliens (Arabs), or as simply people of color (Indians and Pakistanis). The Americanness of these groups is problematic and it raises the following question: do they belong to the United States, or are they part of the United States without belonging to it?[3] That is to say, Americanness in these texts is relationally defined. Rather than emphasizing the consensus of the American experience, its unity and its homogeneity, the emphasis in these texts is on dissensus and difference. Therefore, like the new Americanists, these writers perceive the United States within a transnational context instead of defining the United States in terms of its exceptionalism and isolationism. In the texts I analyze here, Americanness is defined not solely by virtue of national belonging but also along the lines of gender and sexuality, racial, class, religious, and/or ethnic affiliations.

As I mentioned earlier, in my analysis I take into consideration the intertwined history of the United States and Mexico, Cuba, the Arab region, India, and Pakistan. This range of intertwined relationships is described by Radway as "intricate interdependencies" ("What Is in a Name" 10) that rethink nationalism in light of racial, ethnic, cultural, and sexual identities, affiliations, and/or communities, and consequently complicate and reconfigure the relationship between "home" and "abroad," "domestic" and "foreign."

The non-national does not mean having no nation or lack of national affiliations but means problematic national identifications that entail ambivalence and paradox. By non-national subject, I do not mean a subject merely occupying a liminal space between home-country and host-country but rather a subject reconfiguring implications of "here" and "there" within that interstitial space. That is to say, in my discussion of the non-national I problematize the meanings of a liminal subject as simply being "here" and "there" or, alternatively, being neither "here" nor "there." My study complicates occupying an interstitial space between belonging and un-belonging in migrant narratives. In the texts I discuss in this book, there are two moments that help produce the non-national: the moment the main characters realize they do not belong fully as Americans in the United States and the moment they realize that they are no longer perceived as natives of their countries of origin. The non-national moment is not intrinsic to a specific racial group; it rather has multiple meanings, not only among the texts I discuss here but even within the same minority group.

My reading of contemporary narratives by Arab-American, Chicana, South Asian-American, and Cuban-American women writers from the vantage point of an American national space and identity that are constructed within and without the geographical boundaries of the United States situates my study within the new Americanist studies. By the aid of these narratives I will expand the meaning and use of the term non-national. By the same token, by putting Arab-American women's writing in conversation with that by Chicana, South Asian-American, and Cuban-American women writers, I propose Arab-American literature as a potential field that can expand new Americanists' areas of research. Despite the different directions that new Americanist studies has taken—border, hemispheric, or transnational—it still has a lacuna in its lack of reference to Arab-American literature.

Arab-Americans, Who They Are

Michael Suleiman in *Arabs in America: Building a New Future* (1999) observes that the term "'Arab Americans' refers to immigrants to North America from the Arabic-speaking countries[4] of the Middle East and their descendants" (1). Suleiman explains that there have been two waves of Arab migrants, first from 1870 to World War II; and the second from World War II to the present (1). Wars and political changes in the Arab region, specifically since 1948 and the expulsion of Palestinians have impelled Palestinians to move between several countries in the region and have contributed to the rise of Arab migrants to the United States, especially from the Palestinian-occupied territories, Jordan, Syria, and Lebanon (these countries before the British and French colonization were also known as Greater Syria region[5]). Furthermore, the series of wars with Israel that took place in Egypt and Syria in 1956, 1967, and 1973 and their outcomes that led to losing more Palestinian territories and/or more occupation have brought more migrants from Egypt in addition to the migrants from the Greater Syria region. Although the early Arab migrants belong to religious minorities—Christians and Druze—later waves of Arab migrants were from different Muslim sects (Sunni and Shi'a). Like Suleiman's study, most Arab-American Studies in the 1980s and 1990s—especially Eric Hooglund's *Crossing the Waters* (1986); Alixa Naff's *Becoming American* (1993) and *The Arab Americans* (1999); and Evelyn Shakir's *Bint Arab* (1997)—chronicle the history of Arabs' migration to the United States to point out their battles for naturalization and citizenship, the challenge of their being accepted as fellow Americans, as well as the different perceptions of Arab Americans before and after World Wars I and II. Whereas before World War I, Arab Americans thought of themselves as "sojourners," as Suleiman points out (9), staying temporarily in the United States to send money to their native countries; after World War II, the political changes that transpired in the Arab region impelled many Arab migrants to claim a new homeland in the United States without severing political and national ties with their native countries. These changes were also reflected in how Arab migrants in the United States perceive themselves. Because most of the early Arab migrants were from the Greater Syria region, they perceived their identity as Syrians and their affiliations were based on belonging to a certain clan, village, or sect rather than nation. As Suleiman puts it, these identities were "amorphous," "indeterminate," and "shifting from one orientation to another" (7). Nevertheless, as Suleiman elaborates, after the 1967 setback, they perceive their identities as *Arab* to indicate their national and political ties with the Arab region, rather than as solely Syrian (10).

The term Arab Americans,[6] though it includes Arabic-speaking countries, excludes countries such as Turkey and Iran that share Islam as the major religion of their people with the Arab region. Joanna Kadi in *Food for our Grandmothers: Writings by Arab-American and Arab-Canadian Feminists* underscores the limitations of the term Arab Americans: though it affirms affiliations with Arab region, it does not link Arab Americans to larger groups of people of color (Asians and Africans). Furthermore, Kadi points out that using alternative terms such as "people of West Asian/North African descent" and "people of Middle Eastern/North African heritage" are problematic as well. Although the term "people of West Asian/North African descent" includes Middle Eastern and Arab countries, it "will once again make Arabs invisible." In addition, though the term "people of Middle Eastern/North African heritage" includes Iranis, Armanis, and Turks, the term Middle East conjures up the legacy of Western colonizers who have "named the region only as it related to their particular world view" (xviii–xix). Tracing the history of the term, Wael Hassan in *Immigrant Narratives* (2011) points out that the naming of that region as *Middle East* dates back to 1903 when "British and French colonialism drew [the] map [of the region]" (4). The term Arab American replaced the term Syrian, as I have mentioned earlier; it was used more after 1967, when the first Association for Arab-American University Graduates was established. Tanyss Ludescher argues in "From Nostalgia to Critique" (2006) that the Lebanese-Israeli War in 1982, the first Palestinian uprising (*Intifada*) in 1987, and the Gulf War in 1991 further *politicized* the Arab-American community to express their concerns about these political events. In this book, I use the term Arab American to refer to migrants from Arabic-speaking countries, and I select writers from West Asia (Jordan, the West Bank, and Lebanon), and from North Africa (Egypt). Texts by Arab-American writers in my study refer to the political changes that have taken place in the region since the times when the region was under the Ottoman colonization and British mandates to the more contemporary history of the Arab-Israeli Wars. Nevertheless, these texts do not suggest or imply that Arab Americans are homogeneous, but rather reflect the diversity of Arab Americans. For example, Egyptians have cultural roots in the Mediterranean, the African continent, and the Arab region, a diverse heritage that has been strategically used for several political reasons and ideologies. Different political incidents have changed the cultural affiliations of the country. For instance, Gamal Abdel Nasser, after dethroning the monarchy in Egypt, valorized a pan-Arab identity for Egypt and Egyptians. Anwar El-Saddat, who followed Nasser, adopted an Islamic identity that is open to the West. Furthermore, what has contributed to the fluidity of the Egyptian identity is that Egypt has the biggest Christian population in the Middle East. Hence, identifying Egyptians with one single monolithic identity, either as Arab or as Muslim, effaces its diverse cultural legacy. In a similar sense, "Arab-American" identities are defined along shifting political and cultural views, as well as shifting ideologies. Suleiman underlines that Arab Americans "think of themselves in *different* ways at *different* times or in *different* contexts, and they argue . . . for the use of more descriptive categories that recognize different aspects of their background, culture, or physical appearance" (15).

Arab-American literature can be traced to the early twentieth century. In her genealogical study of Arab-American literature, "Arab-American Literature" (1996), Evelyn Shakir distinguishes three phases in Arab-American literature. The early phase before World War I started in the early twentieth century and was pioneered by Ameen Rihani who published the first Arab-American novel, *The Book of Khaled* in 1911, and Khalil Gibran who published his widely known book *The Prophet* in 1923. Rihani and Gibran, as Shakir points out, were the founders of "Mahjar" literary movement in Arabic literature. They initially wrote in Arabic and their works were known in their home countries. They write about the Ottoman colonization, social and political conditions in their home country, introducing European Romantic themes in Arabic literature. Rihani, in his novels, as Shakir points out, serves as "mediator" between Arab and American societies (5). Building on Shakir, Wael Hassan suggests that Arab immigrant writers play a "translational role . . . as interpreters and mediators between their homeland and adoptive countries" (xii). The second generation of Arab-American writers, unlike the first generation, are American-born writers who were known between the 1930s and 1960s. This generation of writers such as Vance Bourjaily and William Peter Blatty have a complex relation to their countries of origin. Literary works such as *Confessions of a Spent Youth* by Bourjaily and *Which Way to Mecca, Jack* by Blatty, both published in the 1960s, depict Lebanese American characters who do not perceive themselves as either Lebanese or Arab. The third phase of Arab-American literature emerged after 1967; that was also the time when Arab migrants began to identify themselves as "Arabs" rather than "Syrians." Hence, the term "Arab" was used as a "statement of solidarity" as Shakir suggests (9), and many post–1967 literary writings were characterized by their "political impulse," as Nabil Alawi puts it in "Arab American Poets: The Politics of Exclusion and Assimilation" (53). This generation of Arab-American writers includes Naomi Shihab Nye, Elmaz Abinader, Sam Hazo, and Sam Hamod, as well as more contemporary writers such as Diana Abu-Jaber, Khaled Mattawa, Joseph Geha, Ramzi Salti, Rabeh Alammdin, and Mohja Kahf. This generation of writers stresses the heterogeneity of Arab Americans in order to dismantle anti-Arab racism in the United States. In this respect, Shakir argues that "[m]ost recently, with ethnicity in fashion at home and Arabs in trouble abroad, Arab Americans have been entering into spirited political debate with one another and with the public at large" (15). I, therefore, situate the Arab-American literary texts in my study within this genealogy to extend Shakir's framework and I argue that contemporary Arab-American literature, especially that written in the past decade with its emphasis on the close relationship between "home" and "abroad," reconfigures both terms and consequently invites the reader to rethink notions of belonging and national affiliations within the United States and the Arab region.

My critical analysis of Arab-American literature is not only in tune with what Shakir proposes but also aims to contribute to

Arab-American literary criticism by suggesting multiple theoretical frameworks and concepts to interpret fiction by Arab American writers. My work resonates with what Steven Salaita suggests in *Arab American Literary Fictions, Cultures, and Politics* (2007) about promoting Arab-American literary criticism: it needs "to be decentered from provincial notions of ethnic atavism and situated instead in comprehensive interethnic dynamics" (6). As I will show in more detail in the coming chapters, my critical analysis of Arab-American texts is comparative, reading Arab-American fiction as well as Chicana, South Asian, and Caribbean writers in the United States. Nevertheless, my approach is not simply comparative to point out similarities among the writers I discuss here, but rather *contrapuntal*, which I borrow from Edward Said. In *Culture and Imperialism* (1994), Said holds that a contrapuntal reading interprets "imperialism as a matter . . . of interdependent histories, overlapping domains," "reading from the center and the peripheries" (259). My contrapuntal reading is attentive to US imperialism and its critique. In other words, contrapuntal reading is a simultaneous awareness of the discourse of imperialism and what it suppresses.

The significance of contrapuntal reading to my study is also that it is an expanded reading that goes beyond one single group or experience. That is to say, my focus on Arab-American narratives does not subsume the experiences of Chicanas, Cubans, or South Asians in the United States. On the contrary, by deploying a contrapuntal framework, this reading strategy does not presume a commonality among the minority groups in my study. It rather conjures up different histories of the usurpation of Mexican territories, economic exploitation of South Asians and Latino/a migrant labor, as well as the unjustified detentions and deportations of individuals of Arab or Middle Eastern descent in the aftermath of 9/11. In that sense, a contrapuntal reading resonates with what Grace Kyungwon Hong and Roderick Ferguson call "strange affinities," making connection through difference possible (18). Hong and Ferguson define "difference," in *Strange Affinities: The Gender and Sexual Politics of Comparative Racialization* (2003), as "the dividing line between valued and devalued, which can cut *within*, as well as *across*, racial groupings" (11 emphasis added). In light of this definition, I argue that reading into the histories of Mexicans, Cubans, South Asians, and Arabs in the United States reveals that different Naturalization Acts, immigration laws, as well as economic ideologies have played a central role in the "racializiation"[7] of these groups differently during different periods in US history. That is to say, "race" has been at the center of exclusionary, or conversely inclusionary, practices toward Arab Americans, Chicanos/as, South Asian Americans, and Cuban Americans; hence creates uneven "(de)valuations" within and among these groups.

For example, Ian Haney-Lopez in *White by Law* (2006) discusses the deportations of Mexicans during the Depression years, even though they were classed "white" and conferred American citizenship since the Guadalupe Hidalgo Treaty between the United States and Mexico in 1848. Similarly, despite their Naturalization as American citizens, individuals of Arab and/or Middle Eastern background were deported, unjustifiably detained, or arrested by virtue of the PATRIOT ACT, which was approved in 2001 after the September 11 attacks. These unjustified practices have been based on the conflation of Arabs, Middle Eastern, and Muslims, or as Nadine Naber puts it, the "racialization of Islam." Naber explains, in "Ambiguous Insiders: an Investigation of Arab American Invisibility" (2000), that Islam has racially marked Arab Americans "on the assumption that all Arabs are Muslims and Islam is a cruel, backward and uncivilized religion" (52). As such, all Arabs are Muslim and all Muslims are potential terrorists.

A different sense of racialization, specifically "racialization of labor,"[8] has been at the center of the experience of South Asian migrants in the United States. South Asians have been denied/allowed entry to and citizenship in the United States by virtue of different labor Acts and laws. Tracing the histories of various Asian groups in the United States, Ronald Takaki in *Strangers from a Different Shore* (1989) points out that Asian Indians were allowed to migrate to the United States as laborers during the late nineteenth century and early twentieth century. In the 1910 census, Asian Indians were classed "white" and conferred citizenship.

Nevertheless, a few years later, according to the "barred zone" law passed in 1917 and the Immigration Act of 1924, Asian Indians were denied entry to the United States and hence became ineligible for Naturalization. Along the same lines, Vijay Prashad, in *The Karma of Brown Folk* (2000), compares the 1965 migration laws to the 1864 Contract Labor Law that allowed larger migration quotas of laborers from South Asia. These laws of "indenture migration" (72), as Prashad concludes, legalized increasing numbers of workers from South and East Asia as cheap workers.

Unlike Mexicans, Arabs, and South Asians in the United States, Cuban migrants were perceived more as "political refugees than settlers" (116), and hence their experiences stand in sharp contrast to other minority groups, especially other Hispanics, as Yehudi Webster points out in *The Racialization of America* (1992). As Webster puts it: "In terms of federal policies, there are Hispanics, and Hispanics" (127). Whereas Mexicans have been perceived as illegal immigrants, encroaching upon the US economic resources, Cubans have been perceived as victims of Fidel Castro's authoritative regime. Webster elaborates that the Cuban migrants who left in the 1950s, fleeing Castro's regime, had unrestricted entry to the United States, and received huge amounts of federal assistance that helped in their establishment of Cuban American economic enclave and financial centers. In other words, early waves of Cuban migrants were "deracialized" and "politically romanticized" (222), as Isabel Molina Guzmán points out in "Disorderly Bodies and Discourses of Latinidad in the Elián González Story" (2010). Unlike other Latinos/as, Cuban migrants of the 1950s and 1960s are a successful model Latina/o minority, many of them are entrepreneurs and politically conservative. Hence, successful Cuban migrants are integrated and socially accepted as "honorary whites" (76), as Coco Fusco holds in *English Is Broken Here: Notes on Cultural Fusion in the Americas* (1995). Fusco elaborates that socially accepted Cubans are paradoxically privileged, yet simultaneously perceived as an ethnic minority. Unlike Cuban migrants of the late 1950s and the 1960s, later waves of Cuban migrants, especially in the 1980s are less educated and poorer. Hence, as

Guzmán explains, recent data collected about the status of Cuban migrants have reflected a change in status from "a *deracialized*, political ethnic enclave associated with whiteness to *racialized* ethnic exiles associated with the brownness of Latinidad" (234 emphasis added). Conjuring up Hong's and Ferguson's terms, these "differences" among Cuban migrants in the United States reveal uneven "(de)valuations" within the same minority group.

In light of this contrapuntal reading of the history of racializing the minority groups in my study, one can infer that racialization of labor and racialization of Islam in the United States unfold the intersectionality of race, class, and religion, and, hence, relational differentiations within multiple hierarchies of power. In addition, one can further highlight that racial categories in the United States are arbitrary and contingent on achieving specific economic, political, or social goals. Furthermore, racial categorization in the United States is inherently flawed and imprecise in the sense that it effaces heterogeneity within the same minority group.

The scope of my study, nevertheless, has its own limitations, especially analyzing multiethnic American writers through the Arab American categorical frame. This analytical framework may overshadow more comprehensive perspectives on the non–Arab American writers and the literary canons they contribute to, such as South Asian, Hispanic, or Latino/a. For example, a more comprehensive analysis would highlight problems with the ethnic categories South Asians, Hispanics, and Latinos/as. Resonating with what I pointed out earlier about the shifting implications as well as the inadequacy of terms like Arab or Middle Eastern, the terms South Asian, Hispanic, and Latino/a share similar problems.

For example, Lavina Dhingra Shankar and Rajini Srikanth, in *A Part, Yet Apart: South Asians in Asian America* (1998), highlight the lack of acknowledgment of South Asian authors within the discipline of Asian American literary studies and hence address problems with the terms South Asian, Asian American, and Postcolonial. As Shankar and Srikanth underline, the term South Asia refers to seven states: India, Pakistan, Bangladesh, Sri Lanka, Nepal, Bhutan, and Maldives. However, individuals who belong to these states identify in national, linguistic, religious, or regional terms rather than as South Asians. By the same token, the term South Asian American literature is the subject of debate among scholars who examine literature produced by South Asian writers. Along these lines, Rajiv Shankar and Lavina Shankar examine the problematics of categorizing South Asian authors as postcolonial or Asian Americans within literary and cultural studies. On the one hand, Lavina Shankar explains, in "The Limits of (South Asian) Names and Labels: Postcolonial or Asian American?" (1998), that the term postcolonial implicitly "obliterates their [South Asians] indigenous histories before European colonization" (53), in the sense that it limits the history of the region to the violent partition of the subcontinent. On the other hand, Rajiv Shankar, in "South Asian Identity in Asian America" (1998), holds that the all-encompassing term, Asian America, effaces the discrepant political and economic histories between the United States and East/Southeast Asia and between the United States and South Asia. As Shankar puts it, "too many differences and divergences keep them [South Asians] 'apart' from the established Asian American (i.e., East/Southeast Asia) identity" (xii–xiii). In a similar sense, Sidhwa's and Laihiri's works provoke similar debates. For instance, Sunil Sharma, in "Pakistani-American Literature" (1996), situates Sidhwa's literary works within the South Asian American literature. He holds that Pakistani-American writers, like Sidhwa, share with other writers of the South Asian diaspora "the memory of colonialism, partition, and the problems of nationalism" (160). Unlike Sidhwa's, Lahiri's works, for some literary critics, are more controversial, as Lavina Dhingra Shankar and Floyd Cheung point out in *Naming Jhumpa Lahiri: Canons and Controversies* (2012), "[i]s she a Bengali, Indian, Asian American, American, or a postcolonial writer?" (vii). Not without its flows, Lahiri's fiction appeals to a wide range of audience, especially in her portrayal of the challenges of migrating and relocating, simultaneously focusing on the Bengali and Indian diaspora. As Lavina Dhingra Shankar puts it, in "Not Too Spicy: Exotic Mistresses of Cultural Translation in the Fiction of Chitra Divakaruni and Jhumpa Lahiri" (2009),

Lahiri's works are "not too spicy" for mainstream American tastes, and hence "accessible to 'non-native' audiences" (27).

Resonating with these theoretical debates about the term South Asian are similar controversies about the use of the terms Hispanics or Latinos/as. On the one hand, except for Brazil, the term Hispanics suggests a linguistic commonality among individuals from Central and South American states that were under the Spanish colonization between the sixteenth and the nineteenth centuries. This category also includes individuals from Spain. However, the term does not account for the different Spanish dialects in the region or the economic disparities within each Hispanic subgroup. As Webster puts it, the term Hispanic does not identify "a mobilized collectivity, a group identity, a regionally concentrated population, or a population with a uniform socioeconomic condition" (115). On the other hand, Latino/a, a geographical term that refers to individuals from Latin America (a vast region that includes North America, Central America, the Caribbean, and South America), does not account for the linguistic, political, and ethnic diversity among different Latin American nationalities. A case in point is the difference between Mexican-Americans and Cuban-Americans. Not only do Mexican-Americans and Cuban-Americans have different histories of (de)racialization in the United States, as I pointed out earlier, but also differences exist within the same minority group due to the different modes of entry throughout different periods of US history. As I mentioned earlier, earlier waves of Cuban migrants were perceived as political victims of Castro's despotism, unlike later waves of migrants, especially in the 1980s, who were perceived as lower, poorer class of migrants. In a similar sense, different terms have been used for self-identification among Mexican-Americans, such as Tejanos/as, Chicanos/as, or mestizos/as. Whereas the term Tejano highlights the history of the geographical borderlands between Texas and Mexico, mestizo emphasizes the mixed heritage of Mexican-American, and Chicano implies the political struggles and histories of Mexicans and Mexican-Americans in the US southwest region. As Webster explains, "Chicanos emerged out of the crucible of

U.S. annexationist and colonial policies in the Southwest" (119). In the coming chapters, my analysis of *The Last Generation* by Cherríe Moraga and *The Agüero Sisters* by Cristina García reflects different, though connected, political and economic concerns for a Chicana author and a Cuban-American author, respectively.

A contrapuntal approach, as I mentioned earlier, does not seek commonalities, it is rather attentive to differences within and across the minority groups I discuss in my study. Thus, the texts I discuss in this book together they offer a broader perspective to analyze notions of the nation, national consciousness, national time, national space, and national belonging. In this way, I use a contrapuntal reading to reinterpret one of the notions essential to the formation of the nation, namely, "imagined community," which Benedict Anderson theorizes in *Imagined Communities: Reflections on the Origins and Spread of Nationalism* (1986). I, therefore, aim at decentralizing monolithic national identifications. In this regard, I am building on the transnational in American Studies that Amy Kaplan proposes in "Violent Belongings": "we must understand how 'America' is a relational, a *comparative* concept" (11 emphasis added). This new understanding of *America* takes in consideration how meanings of *America* have changed historically in different international contexts; it "traces alternative spaces and modes of belonging to collectivities not subsumed by the nation-state...and it reconceives immigration as multidirectional movements" (11–12). My contribution to Arab-American literary criticism is novel in the sense that I do not focus only on common themes in Arab-American fiction or unfold and criticize the stereotype of Arabs in the United States that Arab-American fiction always aims at shattering. I rather interpret Arab-American literary texts within a wider context to re-nuance notions of Americanness in American Studies. In this way, my study differs from Salaita's critical study of Arab-American literature. Salaita's book, *Arab American Literary Fictions*, is the first book-length critical study of Arab-American literature. In his first book as well as in his second critical book, *Modern Arab-American Fiction: A Reader's Guide* (2011), Salaita focuses on common themes in Arab-American literature. He suggests a study of "thematic archetypes" in Arab-American fiction and observes that specific themes recur in most fiction by Arab-American writers, such as the Arab-Israeli conflict, Islamophobia, intersections of race, gender, and national belonging (7–8). Salaita's critical analysis of Arab-American literature is twofold. First, he addresses the absence of scholarly critical studies on Arab-American literature. Second, he situates Arab-American Studies within American Studies or Ethnic Studies rather than Middle Eastern Studies. His reasoning is that "Arab Americans are fundamentally of the United States" and hence "Arab American Studies is best developed within the framework of various American landscapes" (*Modern* 6). Resonating with Salaita is Hassan's study of immigrant narratives by Arab-American writers. As I mentioned earlier, Hassan observes that Arab-American literature *mediates* between the writers' home and host countries, and hence it is "translational literature" (28). In this regard, Hassan examines a range of texts that specifically address the perceptions of Arabs as orientals to conclude that some writers reemphasize the East–West binary, while other writers question and criticize this dichotomy. Hassan points out that for Arab migrants, writing in English has always been "a politically charged translational task, heavily invested in discourses of cultural identity" (29). By the same token, Hassan uses the theory of minor literature by Gilles Deleuze and Felix Guattari to interpret three aspects of Arab immigrant writings: defamiliarizing English, the correlation between individual concerns and collective concerns, and the political nature of Arab-American literature (5–6).

Arab-American Women and Women of Color?

The implications of the non-national in my book are not only tied to meanings of the national in American Studies but are also intertwined with the place of the national and the transnational in narratives by and about women of color. I have chosen to read together different groups of women of color rather than focusing on only one group because they have different histories in the United States and hence different perceptions of the US nation-state, specifically US colonial and imperial history in Central America and the Caribbean Islands, as well as its more contemporary history as the only superpower since the end of the Cold War. My choice of texts raises the following question: can we think transnationally about narratives by Arab-American, Chicana, South Asian-American, and Cuban-American women writers, rather than as only literatures produced within the United States? In other words, can we situate these narratives within the larger context of transnational political, cultural, and economic relations that have created new forms of alliances and communities as well as asymmetries and inequalities?

To unpack this question, I start with questioning the category "women of color" per se. This category is problematic in the sense that it can be too inclusive or too exclusive. On the one hand, encompassing all women based on sharing a "common context of struggle rather than color or racial identifications," as Chandra Talpade Mohanty argues, elides differences among different groups of women, and hence can be too inclusive (qtd. in Stanley 2). On the other hand, limiting the category "women of color" to women who physiognomically have a nonwhite color can be essentialist and hence too exclusive. This category is even more problematic for Arab-American women. This is because, though racially white,[9] they are culturally colored. In other words, Arab-Americans share with other minorities of color in the United States stereotypical representations as well as systematic erasures from American public discourse. However, unlike other groups of color, Arabs are not perceived as a minority, but rather as the "invisible racial/ethnic group," as Nadine Naber puts it in "Ambiguous Insiders" (37), or at best as "foreigners" as Nada Elia points out in "Islamophobia" (156). That is to say, Arab-American women might fit uneasily in the category "women of color."

My doubts resonate with the theoretical concerns that Sandra Soto and Robyn Wiegman raise in "Where in the Transnational World Are U.S. Women of Color" (2005) and *Object Lessons* (2012), respectively, but from different positions. On the one hand, Wiegman explains that the overemphasis on the experience of black women inadvertently repeats white feminisms' exclusions by "similarly subordinating some women's experiences in favor of those of others" (248). On the other hand, Soto raises a different concern. She holds that the category, "women of color," "[has] come to be cited . . . as an undifferentiated monolithic category . . . reifying the homogenizing categories [it] had initially sought to critique" (121). She also points out the limited presence of feminism by American women of color in transnational feminist critiques. In her later article, "Transnational Knowledge Projects and Failing Racial Etiquette" (2008), Soto discusses the example of transnational scholarship in Chicano/a Studies that focuses on how transnational capitalism inscribes the racial formations of Latinos/as in the United States. In this regard, Soto, citing Juan Poblete, underlines the colonial history of the United States in Mexico and Puerto Rico; and the significant role American imperialism has played in bringing more migrants from Latin America to the United States (69). I share Wiegman's and Soto's concerns and I add that despite acknowledging Arab-American communities and the increasing interest in including Arab-American women writers in anthologies about women of color,[10] yet there is need for more work to be done on Arab-American narratives from a transnational perspective. Thus, I expand Soto's discussion of the imperial history of the United States in Central America, specifically in Chapter 2, where I analyze *West of the Jordan* by Laila Halaby and *The Last Generation* by Cherríe Moraga. I interpret both texts through the lens of the political and economic relationships between the United States and the Palestinian-occupied territories and the United States and Mexico.

The ambiguous positionality of Arab-American women as "not quite white" and "not quite women of color"—as Rabab Abdulhadi, Evelyn Alsultany, and Nadine Naber point out in their introduction to Arab and Arab-American feminisms (xxxiv)—raises another concern. That is, the reception of Arab-American writers in general and women writers in particular. Hassan points out that, specifically, narratives of Arab women's oppression "are eagerly received" (36). On the one hand, these narratives reemphasize the stereotype of Arab women as intrinsically victims of Arab male tyranny; on the other hand, they justify US policies of dominating the Arab region (or more largely the Middle East) to save Arab women.[11] Similarly, Abdulhadi, Alsultany, and Naber hold that within the US feminists' discourse, Arab-American feminists feel the pressure of responding to and correcting stereotypes of the veil, the harem, and female circumcision (xxxvi). That is to say, mainstream feminist discourse limits the parameters of Arab-American feminist writings to address the *orientalist* stereotypes of Arab women rather than conceptualize and theorize their lived experiences, and hence preempts Arab-American feminist discourse.

Another factor that has contributed to the marginalization of Arab-American writings is the assumption that it is necessarily anti-Israel as well as the misrecognition of the Arab-Israeli conflict as Arab-Jewish conflict.[12] This reasoning finds support in the conflation of Arabs and Muslims and the stereotypical perception of Muslims as terrorists or potential terrorists. Hassan cites the example of Gregory Orfalea's memoir that was denied publication because of the writer's Arab identity, even though it portrays intermarriages between Muslims and Jews in 1936 in Palestine (36–37).

In this study, by pairing Arab-American women writers with Chicana, South Asian- American, and Cuban American women writers, my goal is to re-nuance the perception of the nation and the national from their point of view vis-à-vis other minority groups in the United States. To that end, I analyze texts in this study through the critical lens of what Kyla Wazana Tompkins, in "History's Traces," calls "strategic reading—and writing—practices" (126). Tompkins explains that strategic reading and writing practices are tied to "transnational politics of location"; they take "displacement as a starting point for affiliation and alliance with other displaced peoples" (137). These affiliations and alliances do not lie in one singular community. Building on Tompkins, I elucidate strategies that emerge in the narratives I analyze in the following chapters, which evoke what I call the *non-national.*

Non-national Sites

I start by reconfiguring Anderson's concept of "imagined communities." This reconfiguration emanates from my analysis of the formation of ethnic/national communities in *The Language of Baklava* by Diana Abu-Jaber and in *An American Brat* by Bapsi Sidhwa. I chose these two texts specifically because they both focus on coming to America and/or growing up in America themes. Ethnic/national loyalties in both narratives are formed within and outside the geographical boundaries of the United States. In both texts, the Althusserian moment of "interpellation," in which the individual is hailed as a subject, is interrupted because of the ethnic/religious identities of the main characters: Diana in *The Language of Baklava* and Feroza in *An American Brat*. I argue that an interrupted moment of interpellation is a non-national moment in which the national consciousness of the subject is reshaped. The understanding of national consciousness that emerges in my analysis of *The Language of Baklava* and *An American Brat* is not bounded by essentialist notions of national belonging or ethnic identification. In *The Language of Baklava*, food is a trope around which communities are formed as well as negotiated to symbolically stand for situated affiliations rather than the national or ethnic. In a similar sense, in *An American Brat*, the moments of estrangement that the main character encounters in both Pakistan and the United States re-nuance the definition of national consciousness and create a new sense of imagined community formed through estrangement rather than from shared feelings of belonging.

In Chapters 2 and 3, I propose a new corollary to understand Anderson's "imagined community," namely, if communities are not nation-centered and formed across geographical territories, the time and space that the non-national subject occupies are not nation-centered either. In the second chapter, I discuss *West of the Jordan* by Laila Halaby and *The Last Generation* by Cherríe Moraga to elucidate meanings of non-national temporality in the former and queer temporality in the latter. The lineage of a national identity transmitted from parents to children is problematic in both narratives because of displacements and migrations caused by occupation (as in *West of the Jordan*) or conquest (as in *The Last Generation*). Both texts bring together discrepant geographies and histories and make them comparable—the United States and West Bank in the former, and the United States and precolonial America in the latter—and hence complicate the perception of US national time. By the same token, I argue that non-national and queer temporalities make it possible to retell the story of the US nation-state from the perspectives of the displaced and the exiled.

To that end, I build on Ernst Bloch's theory of "contemporaneous non-contemporaneity" in *Heritages of Our Time* (1935) and on Julia Kristeva's concept of "monumental temporality" in "Women's Time" (1981). I situate non-national time in Bloch's concept "contemporaneous non-contemporaneity" to argue that non-national time is a paradoxical, interstitial time. Non-national time is formed specifically at a conflictual point, in which the main character, Hala, in *West of the Jordan*, has a contentious relationship with her father that unfolds paradoxical moments in which she looks back at her life in Jordan to recollect past stories and looks forward to her future in the United States. Hence, the narrative structure of *West of the Jordan* has a paradoxical temporality in which the progression of the story line is interrupted by cycles of recollecting stories from Jordan and the West Bank, which, in turn, interrupt her "coming to America" story. The communities imagined in these stories are not only imagined transnationally but they are also dispersed between two different time zones—United States and the Arab region—hence interrupting linear understandings of national time. Similarly, I argue that Moraga's revisited story of Aztlán, "Queer Aztlán," is rooted in queer temporality. Building on Kristeva's monumental temporality, which she defines as "infinite," I suggest that Moraga's queer temporality puts queer motherhood at the center to replace the conventional symbolism of the nation as a patriarchal heteronormative national family, and consequently displaces the patriarchal genealogy of the nation.

My third chapter elaborates the notion of "imagined (transnational) communities" and I argue that the notion of homeland as a national space is reimagined to unfold a problematic perception of the United States as a new homeland in migrant narratives. I draw on Doreen Massey's distinction between space and place in *Space, Place, and Gender* (1994) to analyze the reconfiguration of home(land) as a non-national space in *The Time between Places* by Pauline Kaldas and *The Namesake* by Jhumpa Lahiri. Both texts problematize the conventional perception of home-country as solely a place of roots and belonging, because characters traveling between their home-countries and the United States do not feel that they fully belong to either their home-countries or their host-country. Massey holds that *space* is a dynamic notion formed and shaped by social relations and *place* is a particular moment in these relations. Although both texts portray the allure of the American dream, my discussion reveals that there is a perpetual tension between feelings of settlement and unsettlement in both the United States and the native countries of the main characters. Within these stories of migration there is a particular moment of dislocation, which I call non-national, evoked by feelings of un-belonging and unfulfilling social relations. The conventional perception of the United States as the land of freedom and opportunities is problematized in *The Time between Places* and *The Namesake* when some of the main characters in each story are successful, while others are unable to find their place in the land of opportunities.

In Chapter 4 I reopen the theoretical debates about national allegories initiated by Fredric Jameson in "Third World Literature in the Era of Multinational Capitalism" (1986). Jameson observes that literatures produced in Third World nations in the postindependence era are open-ended and allegorical. This narrative form is tied to the political crises that many Third World nations underwent: specifically their disappointment in fulfilling the national goals of independence because of a dominating, expanding capitalism. My discussion of *The Agüero Sisters* by Cristina García and *The Night Counter* by Alia Yunis in this chapter suggests that while these narratives may reflect national and political concerns, such as socialism in Cuba, global capitalism, and the aftermath of 9/11 in the United States, these concerns are not geographically bounded within a singular nation or founded in a first-world—third-world binary, but rather unfold an intertwined political and economic history between the United States and Cuba (as in *The Agüero Sisters*) or the United States and the Arab region (as in *The Night Counter*).

I draw on Jameson's theory of national allegory, especially his definition of allegory as a "signifying process" that generates complex meanings and messages. However, I extend Jameson's theory by problematizing the notion of the nation in his argument. In both *The Agüero Sisters* and *The Night Counter*, economic relations between the United States and the Caribbean Islands and the Arab region transcend geographical boundaries, yet they are uneven and complicate the relationship between individuals and their national affiliations. The non-national, as it emerges in my study, contests singular identifications and opens up new forms of political affiliations, yet it does not lend itself to allegory. Therefore, I suggest that *The Agüero Sisters* and *The Night Counter* can be read as national allegories, reflecting national concerns, and can be also read as transnational allegories, specifically in light of global capitalism.

CHAPTER 1

The Non-national Subject in *The Language of Baklava* and *An American Brat*

In the different notions of Americanness that circulated at the turn of the twentieth century, there were tensions between maintaining ethnic identifications, valorizing individualism and liberty as eminent elements of an American national consciousness, as well as a third definition of America as a transnationality. James Bryce holds, in *The American Commonwealth* (1888), that "individualism, the love of enterprise, and the pride in personal freedom, have been deemed by Americans not only their choicest, but their peculiar and exclusive possessions" (419–420). Individualism and freedom are not only fundamental in definitions of Americanism but also the core principles of national identification that unite Americans, as opposed to unity based on shared ethnic/religious roots.[1] However, Randolph Bourne, in "Trans-national America" (1916), argues against narrowing the conception of Americanness to the Anglo-Saxon traditions. Bourne holds that, "America is coming to be, not a nationality but a trans-nationality, a weaving back and forth, with the other lands, of many threads of all sizes and colors" (121). The escalation of different waves of immigrants to the United States from the late nineteenth century heightened the feeling of urgency to find an ideal that could function as the ground for American national identification; individualism served that goal. It is only by adopting ideals of personal freedom that the immigrant is Americanized and will "possess *the national consciousness of an American*" as Louis Brandeis maintains (640 emphasis added). To that end, Woodrow Wilson, in his speech to naturalized citizens in 1915, underlined that "[a] man who thinks of himself as belonging to a particular national group in America *has not yet become an American*" (American Presidency Project, emphasis added). These contrary perceptions of Americanness imply that there are two kinds of consciousness and identifications that are set in contrast: one is national and inclusive, based on the ideal of individualism, which unites *all* Americans; and the other is ethnic and exclusive, relying on group belonging, and hence is deemed not quite American. However, "ethnic" is not a self-identity, it is rather a political construction that has developed historically to grant, or conversely, deny specific ethnic groups their political rights based on the binary of white and not-white. Ronald Takaki in "Reflections on Racial Patterns in America" (2008) critically examines the racial patterns in the United States since the eighteenth century that have denied nonwhite individuals (e.g., blacks, Native-Americans, and nonwhite immigrants) rights of citizenship and political franchise. In other words, the term "ethnic" has been used to signify nonwhite individuals and to position ethnic identifications in opposition to national identifications. *The Language of Baklava* (by Diana Abu-Jaber) and *An American Brat* (by Bapsi Sidhwa) revolve around the struggle that the female protagonist in each text undergoes, growing up in America, trying to become "American" or to be accepted as an "American," and acknowledging her "ethnic" loyalties.

Diana in *The Language of Baklava* is born in the United States—her father is Jordanian[2] and her mother is American—yet wants to define herself as Jordanian and American rather than only American or only Jordanian. Her counterpart in *An American Brat*, Feroza, is sent by her parents in Pakistan to visit the United States. Feroza wants to perceive herself as Parsee and Pakistani as well as American. These desired terms of self-definition are contradistinctive to the definition of Americanness founded on discarding group belongings or ethnic identifications. In Woodrow Wilson's definition of the American national consciousness, Diana and Feroza have *not yet* become American because they maintain their "ethnic"/religious belonging. However, both characters do not see a contradiction between adopting ideals of individualism and personal freedom on the one hand and belonging to their ethnic/religious groups on the other. I define the *not yet* in light of Slavoj Zizek's theory of "transubstantiation" in "Multiculturalism, Or, the Cultural Logic of Multinational Capitalism" (1997). Zizek holds that "transubstantiation" is a process by means of which the tension between an individual's primary "particular ethnic identity" and his/her "universal identity as a member of a Nation-State is surpassed," when the individual recognizes "the substance of [his/her] being in a secondary community [Nation-State]" and cuts links with his/her "primordial 'organic' community [family and local community]" (42, 41). Building on the tension that Zizek describes, I argue that the *not yet* is a disrupted, incomplete process of shifting from the "particular ethnic identity" to the "universal identity as a member of a Nation-State." *Not yet* American is also a cumulative identity that is formed by constantly deferring singular identifications; hence, challenges both ethnic self-enclosures and essentialist national belongings. I expand my definition of the *not yet* to argue that the *not yet* American is the non-national subject. The non-national subject's affiliations and identifications are beyond geographical boundaries of nation-state, and are constructed and severed situationally. This subject's sense of belonging is formed at the disjuncture between the ideal of individualism (eminent to American national identification) and ethnic/religious group belonging. This disjuncture creates sites of ambivalence in which I position the struggles of both Diana in *The Language of Baklava* and Feroza in *An American Brat* with notions of choice and individualism.

Ideals of individualism and freedom of choice are central to American liberalism that underwrites the image of an exceptional America. Bonnie Honig in *Democracy and the Foreigner* (2001) points out the problematic status of the immigrant in the exceptionalist accounts of American democracy, as both reinvigorating American exceptionalism, and also posing a threat to its core values of choice, individualism, just economy, and sense of community. Along the same lines, Ali Behdad in "Nationalism and Immigration to the United States" (1997) underlines the nation's ambivalence toward immigrants: while the discourse of exceptionalism valorizes the United States as a nation of immigrants, the immigrants are perceived as the other to be controlled and normalized. I therefore argue that non-national consciousness (formed at the disjuncture between ideals of individualism and group belonging) allows a sense of "imagined community" that is not nation-centered and is not confined to the geographical boundaries of one nation (United States, Jordan, or Pakistan). In other words, I recall Benedict Anderson's "Imagined Communities: Reflections on the Origin and Spread of Nationalism" (1991) to venture a redefinition of his central concept to be "imagined (transnational) communities."

Anderson holds that the formations of modern national communities (which he traces to the early nineteenth century) are "boundary-oriented and horizontal" (15). I repurpose Bourne's definition of America's transnationality as "a weaving back and forth, with the other lands, of many threads of all sizes and colors" to argue that communities and comradeships imagined in *The Language of Baklava* and in *An American Brat* neither are boundary-oriented nor are they necessarily horizontal; they are formed in both Jordan and the United States (as in *The Language of Baklava*) or in both Pakistan and the United States (as in *An American Brat*). The communities and the affiliations that are formed in *The Language of Baklava* are formed around food. That is to say, food plays the same role that the novel or newspaper has played in enabling people who do not know each other to imagine the presence of other individuals like themselves but not necessarily residing in the same territories. In my analysis of *The Language of Baklava*, food is a trope and a sign that I read semiotically to examine the communities that are formed around it. Likewise, the communities that are formed in *An American Brat* are not confined within geographical boundaries but rather formed around religious, political, and/or state ideologies[3] (such as individualism, capitalism, and Islamization) in Pakistan and in the United States. I use the term "ideology" in Raymond Williams's sense: a set of ideas internalized by individuals. As I will clarify in my analysis of *An American Brat*, different ideologies create different senses of belonging and un-belonging.

Imagined (transnational) communities are transnational in two senses: they are formed beyond the geographical boundaries of a single nation and are not nation-centered. These new formations are evoked by disruptive moments, which I term non-national because they problematize the individual's sense of national identification.

Abu-Jaber's book is a "food memoir," as she describes it in an interview with Robin Field (224). This culinary narrative takes food as the primary means adopted by the characters to define their national-ethnic identities in both Jordan and the United States. *The Language of Baklava* is the memoir of the writer, who grows up in the United States; her father is Jordanian and her mother is American. Her father struggles with homesickness, perceives himself as only Jordanian (never defines himself as American for the major part of the memoir); demands that his daughters should define themselves in the same way; and cooks to keep alive the Jordanian heritage in himself, and by extension in his daughters. *An American Brat* is another growing up story of a Pakistani girl—Feroza—who is sent to the United States by her parents after the Islamic regime took over in Pakistan in 1978 and imposed its rigid rules on all Pakistanis, especially women, whether they were Muslims or not. Feroza is a Parsee who has to encounter the hardships of living under strict Islamic laws as well as the difficulties of adjusting to a different culture in the United States. That is to say, *An American Brat* interweaves the growing-up plot with the coming-to-America narrative. Rosemary Marangoly George, in "But that Was in Another Country: Girlhood and the Contemporary 'Coming to America' Narrative" (1998) holds that coming-of-age/going-to-America stories "partake of and propagate a developmental narrative in which Americanization and the very American ideal of individualization plays a vital part in shaping and establishing the full adult" (140). The developmental story in *The Language of Baklava* and *An American Brat* differs from the generic coming-to-America story because the female protagonist grows up in both the United States and Jordan (Diana) or the United States and Pakistan (Feroza). Therefore, I argue that Diana and Feroza undergo a complex process of *interpellation*[4] that shapes their relationship to the United States and to Jordan/Pakistan. On the one hand, the parents of each protagonist emphasize a cultural difference between the United States and their parents' native country.[5] On the other hand, Diana and Feroza struggle with "domestic Americanization," which is "making *one* people [American] out of many" as Donald Pease holds in "C.L.R. James Moby-Dick and the Emergence of Transnational American Studies" (2002, 151 emphasis added). In other words, both protagonists do not resolve the tension between their "particular ethnic identity" and their "universal identity as member[s] of a Nation-State." The non-national consciousness of the *not yet* American subject is shaped by maneuvering, challenging, and/or succumbing to the First World–Third World binary and "domestic Americanization."

The Language of Baklava

As I mentioned earlier, *The Language of Baklava* is a "food memoir." Published in 2005, it has not received much critical attention, unlike the novel that preceded it in 2003—*Crescent*—where food is also the central metaphor. Carol Bardenstein in "Beyond Univocal Baklava: Deconstructing Food-as-Ethnicity and the Ideology of Homeland in Diana Abu-Jaber's *The Language of Baklava*" (2010) attributes this lack of critical interest in Abu-Jaber's text to its genre, namely, cookbook-memoir.[6] Bardenstein holds that this genre is associated with nostalgia and is conservative in its representation of ethnicity in the US context (162). Most reviews of the memoir seem to validate Bardenstein's point of view, especially the ones that underline one or two of the following themes: emphasizing the importance of food as a key connection to homeland, especially in immigrant literature; seeking identity through the foods of childhood; bridging the difference between the Arab and the American worlds by means of food; and bearing the pressure of being a good Arab girl.[7] Though these reviews point out the central presence of food as a metaphor to convey the main argument(s) in the book—homesickness and bridging cultural differences—they do not address the complexity of interpreting food as a cultural habit that goes beyond the simple correlation between "what we eat" and "who we are"; as Donna Gabaccia reminds us, "[i]f we are what we eat, who are we?" (9).

Gabaccia, a historian of American eating habits, in her often-cited study about specific diets and different historical periods in the United States—*We Are What We Eat: Ethnic Food and the Making of Americans* (2000)—concludes that colonial America and early years of US independence are more hybrid in culinary terms unlike the more conservative nineteenth and twentieth centuries. She also shows that "American" social values were perpetuated in the era of American national cuisine (between 1870s and 1920s), which witnessed movements by prohibitionists and food reformers aimed at creating an American national cuisine and Americanizing the new immigrants by rejecting foreign ways of eating habits and cooking. That era, as Gabaccia explains, emphasized the puritan New England cuisine as the "scientific, modern, and *patriotic diet*" (122 emphasis added) "necessary for *national strength and health*" (124 emphasis added) and, therefore, appropriate for "*American citizens*" (125 emphasis added). The direct correlation between Puritan New England/white cuisine, patriotism, and national strength on the one hand, and American citizenship on the other hand, resonates with patterns of citizenship in the United States that are founded on the white and not-white binary. As a corollary, foreign foodways were unhealthy, unmodern, unpatriotic, and by extension un-American. Though these movements waned and American cuisine was redefined regionally—rather than nationally—to reflect the US "culinary cultural pluralism"—as Gabaccia puts it (123)—a distinction still remains between "American" cuisine and "ethnic" cuisine.

Categorizing the foodways of immigrant groups as "ethnic" is an arbitrary assumption of an ethnic unity that does not really exist. The inference of this arbitrary assumption is that there are two cuisines in the United States: one defined regionally, not ethnically marked, and the other defined "ethnically," pertaining to particular groups. Whereas in the former the "regional" does not overshadow "national" belonging, in the latter the "ethnic" and the "national" problematize each other.

Is naturalization into American citizenship "empowerment or marginalization" (221)? R. Radhakrishnan raises this concern in the context of the Indian diaspora in the United States, and I see it as relevant to the Arab-American community as well. He points out that naturalization makes the ethnic subject subordinate to its nationalized American status and the "ethnic" becomes a mere qualifier rather than a cultural or political identity (221–222). The ethnic-racial-national identity of the Arab-American community in the United States complicates Radhakrishnan's question even more. Racially categorized as "white," Arab-Americans, however, do not have "legal position within the spectrum of minority cultures from which [they] can legally articulate [their] communal concerns about discrimination," as Carol Fadda-Conrey says in "Arab American Literature in the Ethnic Borderland: Cultural Intersections in Diana Abu-Jaber's *Crescent*" (2006, 188). Categorized as "non-European whites," Arab-Americans are not perceived as the privileged "white" race. Lisa Suhair Majaj in "Arab-Americans and the Meanings of Race" (2000) describes Arab Americans' racial categorization as "honorary" whiteness (320). "Non-European white" is not a specific qualifier for Arab Americans. This renders Arab Americans "the Most Invisibles of the Invisibles," as Joanna Kadi holds in *Food for Our Grandmothers: Writings by Arab-American and Arab-Canadian Feminists* (1994, xix) and as Abu-Jaber in an interview citing Edward Said says: "the Arab was the last ethnicity that it was okay to denigrate and to be openly racist about" (219). One can infer that the national status of Arab Americans as "Americans" is almost overlooked because of their Arab heritage and always conflated with being Muslims,[8] although not all Arabs are Muslims and not all Muslims come from the Arab countries. Food and restaurants are other means of defining Arab Americans, especially known for dishes like tabouleh, hummus, baba ghanouj, and falafel. However, restaurants that primarily offer these dishes are not categorized as specifically Arab-American. They are, rather, categorized as "Middle Eastern," an equally ambiguous category for two reasons. First, it overlooks cultural diversity within the Arab region. The aforementioned list of dishes is more specific to the Levant region rather than to the Arab-Gulf states (like Saudi Arabia, Kuwait, Qatar, or United Arab Emirates) or to the Arab countries in North Africa (like Algeria, Tunisia, and Morocco). Second, the category "Middle East" includes a wide range of countries that do not share Arabic as the native tongue, like Turkey, Israel, and Iran.

I argue that food and recipes in Abu-Jaber's *The Language of Baklava* interrupt the assimilation narrative, reveal the complexity of defining Arab Americans' ethnic-national identity through cooking as improvisation, and produce a *not yet* American subject. The complexity of the ethnic-national identity of an Arab-American subject and its interpellation in the United States is right there on the very first page of Abu-Jaber's memoir. Diana's features are a mix of a "white" American mother (Irish, German, Swiss, Dutch) and an Arab father (Jordanian). Diana, mostly, takes after her mother: light skin/eye/hair color. In addition, her first name—Diana—is quite familiar in the United States and is easily pronounced. However, her last name—Abu-Jaber—sounds odd, compared to her "Western" features and her first name. In the opening scene of the memoir, Diana is a child hosted, among others, in a TV show for children. The broadcaster, having difficulties pronouncing what seems to him "foreign" names—such as Farouq, Ibtisam, Jaipur, Matussem—feels relieved when he comes across Diana's name. However, the broadcaster "crashes into" Diana's last name, trying to articulate it: "Ub-abb-yuh-yoo-jo-jee-buh-ha-ree-rah . . . " (sic, 3). When asked, "what kind of a last name is that," Diana sarcastically answers: "'*English, you silly*!' into his microphone" (3). This first scene of (mis)communication between

Diana and the broadcaster and the laughter it evokes sets the tone for the rest of the memoir regarding her entry as a subject in both the United States and Jordan. Diana's dilemma throughout the memoir is whether she is perceived as "ethnic" or "American" (in the United States) or, conversely, as "American" or "Jordanian" (in Jordan). Althusser holds that interpellation is central to the subject *of* ideology (i.e., a subject that is hailed into ideology) who hence becomes "subject *to* ideology, having to obey its rules and laws and behave as that ideology dictates" (135 emphasis added). Ideology functions through four apparatuses, which Althusser calls "realities which present themselves to the immediate observer in the form of distinct and specialized institutions" (143). These apparatuses are: family, education, religion, and mass media (143). "The Brown DeMone"[9] show that airs on national television and hosts and hails Diana, Farouq, Ibtisam, Jaipur, Matussem is in this scene an ideological apparatus that assumes acknowledgment of diversity as the foundation of "Americanness." Nevertheless, the broadcaster's confusion about names that do not sound *English* unfolds a counter-ideology, interpellating Diana as a subject who belongs to a "particular ethnic identity." Both Diana and the broadcaster in this scene are trapped in a moment of incomplete interpellation in which the conversation stops: he is confused by her last name, and she is surprised at his confusion. However, the scene concludes with the broadcaster and Diana laughing, "but at two *different* jokes" (4 emphasis added). Un-understood, Diana is the *not yet* American subject who throughout the memoir will feel an outsider both in the United States and in Jordan. Every time Diana leaves the United States for Jordan or Jordan for the United States, there are multiple scenes in which she occupies a place as an American or a Jordanian subject but *not yet.*

In this first scene Diana's features and first name categorize her as—literally—"white," while her last name does not. Therefore, while Diana sees herself to be like the rest of the children in the show, the broadcaster sees her cultural belonging(s) to be more problematic. On the one hand, although Diana's first name sounds familiar to the broadcaster, her last name does not. On the other hand, Diana does not see her name or herself to be in any way different from "Farouq, Ibtissam, Jaipur, Matussem." Diana saying that her name is "English," into the microphone on national TV, is like an announcement made to emphasize belonging and ask for an acknowledgment of that belonging. It stands for bringing Arab Americans into the public sphere and making them visible in the mainstream Anglo-American media discourse. It is also an attempt to "naturalize" Arab Americans into American citizenship, which can work either to help acknowledge the presence of Arab Americans who are both Americans and Arabs (living among other Americans in the United States) or to paradoxically "minoritize" the ethnic/racial identity of Arabs. However, if Arabs are "white" but not "precisely white"—as Diana's grandmother describes Diana's father (90)—the "white" becomes a problematic signifier for Arab Americans; it does not empower Arabs to be perceived as a white community because their racial identity is overlooked and conflated with Islam. Diana, a mix of a not "precisely white" Jordanian father and a white European-American mother, is not a happy hybrid mix. When Diana's father relocates the whole family in Jordan, Diana starts to feel her mother's and her own difference from the rest of the family members. She says: "I sense a deep weirdness about my own existence in the world. How could these two people have ever found each other? How could I have ever come to be?" (64) Diana starts questioning her mixed heritage only when the family moves to Jordan. Diana's confusion about where she belongs in this scene stands in contrast to her assertive answer to the broadcaster in the first page of the memoir about her last name or to her mother when she describes herself as simply born in Syracuse: "I was born into the Syracuse world. I have no inkling of what other worlds are like" (20). What Diana experiences as a child in Syracuse is her father's feelings of homesickness and nostalgia, not hers. Therefore, when her mother tells her that they are moving to live in Jordan, Diana simply asks: "Don't we live in America?" (30) This means that Diana has a firsthand experience of belonging to the United States and a second-hand experience (through her father) of belonging to Jordan.

The opening of the memoir diverges from the coming-of-age/coming-to-America plot, which usually begins with "a brief chapter that dramatically cuts between origin (the past country of childhood) and final destination [present location in the U.S.]," as George points out (144). In Diana's story, the "past country of childhood" is the United States rather than Jordan and the final destination can be the United States or Jordan. Diana's father, Bud, insists that his daughters will *grow up* in Jordan, unlike his acquaintances who are keen that their children grow up and become "American." Diana grows up in both Jordan and the United States. Thus, the linearity of the growing-up/coming-to-American story is disrupted. Another twist in the growing-up story in *The Language of Baklava* is that the first chapter in the memoir is titled "Raising an Arab Father in America." That is to say, the memoir is not only about Diana growing up in the United States/Jordan but also about her father Bud who, throughout the novel, pursues the American Dream of success and having his own restaurant. On his way back from Jordan to the United States, Bud admits to an old friend that he is "American" and "is no longer—not entirely—Jordanian" (326). Like Diana's story, there is no linear progression of Bud's story. Bud pursues his dream restaurant in both the United States and Jordan. An origin point and a past country of childhood are mediated in the memoir through Bud's nostalgia and memories about Jordan. Bud's story is mirrored in the chapter titled "Immigrants' Kids." The whole chapter is dedicated to fathers who are haunted by the past and insist on teaching their children that they are not "Americans." Olga's father, Basilovich, Russian-Polish-Ukrainian-Jewish, struggles with severe homesickness, is unable to overcome his feelings of loss, and ultimately commits suicide. Bud's story with no chapter of origin and Basilovich's sad story of maladjustment are glimpses of resistance to the generic coming-to-America story that perpetuates the notion of the United States as the final destination.

Growing up in Jordan, for Diana, is complex because it is mediated by her perception of herself as "American." This complexity is encountered in two incidents: one in Jordan, at age seven, when she eats "native"/Jordanian food but tries with her mother to cook pancakes, and in a second incident when she is thirteen and bakes baklava with her aunt Aya (from Jordan) in the United States.

Echoing the first scene of Diana's interpellation in the United States where her last name cannot be pronounced because it does not sound American/English, in Jordan, Diana's first name is confusing to some of her acquaintances and is pronounced "dee-ahna" (58) because "Diana" does not sound Arabic. Her acquaintances in Jordan attempt to make her first name sound more Arabic; this is, implicitly, an attempt to make Diana a "native" of Jordan. In contrast, Bennett (Diana's British friend in Jordan) attempts to *de-nativize/de-Jordanize* her by scolding her for eating "native [Jordanian] food." Bennett tells Diana: "You don't *belong* with them! You *know* that. You *know* that. The sort you are *belongs* with the sort I am. Like *belongs* with like.... No in-betweens. The world isn't meant for in-betweens, it isn't done. You *know* that." Bennett impatiently sums up his argument with Diana, repeating the same idea: "They *belong* with their own kind. You with me, they with them.... No in-betweens. It's not allowed" (49 emphasis added). Bennett draws an affiliation with Diana on the basis of his perception of himself and of Diana as nonnatives of Jordan: they do not belong to Jordan and therefore have a shared knowledge of their un-belonging. At the other end of the spectrum, Munira—the Bedouin woman who works for Diana's family in Jordan—says to Diana: "this is [Jordan] where you *belong*" (67 emphasis added). Bennett believes Diana should not belong to Jordan by virtue of her features that are closer to his British features than most Jordanians, and Munira believes that Diana should belong to Jordan by virtue of having a Jordanian father. Though Diana starts to acquire some Jordanian Arabic and adjust to Jordanian food—both are acquired through habituation—she craves and misses American ice cream, pancakes, and hot chocolate.

Attempting to cook pancakes in Jordan, Diana and her mother improvise with some ingredients, replacing syrup with honey. Though the ingredients are different, "it doesn't matter too much because we call them pancakes and they look a bit like pancakes."

Munira likes eating them and the neighbors like calling them "burnt American flat food" and eat them side by side with local accompaniments: sesame seeds, fragrant mint, yogurt, cheese, olives, tomatoes, eggs, and pistachios (33). Thus, a new dish comes into existence—burnt American flat bread—and new affiliations are formed around that dish among the "American" mother, the Jordanian "natives," and the "in-between" daughter. Adding the marker "burnt" to signify the peculiar taste of this new dish *ethnicizes* American pancakes in Jordan by giving them a dark color. This incident blurs the demarcations between "ethnic" and "American" as absolute signifiers with intrinsic meanings attached to them. Thus, Diana wonders: "Am I still an American?…it seems like a kind of unbecoming or rebecoming" (58). The question about Diana's belonging has changed to a question of "unbecoming or rebecoming." Is it un-becoming "American" or rebecoming "ethnic" or "American" in Jordan? As I have pointed out earlier, according to Woodrow Wilson, possessing an American national consciousness is incongruent with belonging to national groups. Abu-Jaber's text rewrites this formula by destabilizing the notion of "origin" as a core principle of belonging to national groups so that belonging to a national group and becoming American are not antithetical. In *The Language of Baklava*, Abu-Jaber gives up the notion of "origin" in favor of "becoming." Becoming implies transformation. In actuality, Jordan the land of origin for Jordanian "native" food is a land and a name constructed over different periods of time in history. Before it was created in 1921 it was called "Transjordan." "There were Assyrians, Nabataeans, Romans, Alexander the so-called Great, Persians, Jews, Christians, Muslims, Crusaders, Mamluks, Turks…you can't imagine the comings and goings!" (128) Diana's father insists that despite these numerous comings and goings there have always been the same families of Bedouins living in Jordan. However, ironically, Bedouins are known to be travelers, moving from one place to another, settling and resettling. In the same sense, the *authentically* Arab dessert—baklava—does not have a place of origin. It can be Turkish (191), Greek (185), or Arab (189). In short, "everyone invented baklava"

(189). At the age of 13, back in the United States, Diana is rebellious; she wants to identify herself and her family as "Americans" and hence rejects "Arabic" food. Nevertheless, what Diana initially rejects—baklava—because of its Arab origin, she ultimately ends up eating because of its nonspecific origin. Thus, the "language" of "baklava" is a shifting signifier that shuttles between the transnational and the ethnic, transcending geographical boundaries to form new communities and affiliations.

Denationalizing food, as the example of baklava shows, echoes denationalization of American fast food chains that have spread globally. However, the decentralization of the origin of baklava is different from the sense of decentralization latent in fast food chains. Recipes, although set as instructions for cooking a specific meal and meant to describe a specific dish from a specific region, are open to improvisation (e.g., cooking pancakes in Jordan). In contrast, although the flavors of fast food meals are domesticated, it must appear to fit the standard taste of the same food anywhere in the world. In addition, a recipe reflects a food practice that emphasizes the process of cooking and preparing food, whereas fast food conceals the process of cooking from the consumer. In the memoir, cooking and recipes are personalized, unlike ready-to-eat food that is represented as a standardized commodity. The recipes in the memoir constitute a multivalent narrative that implies a set of images, tastes, choices, and values. They are personalized in two senses: first, they are situated within the context of the story to evoke a related emotional moment. For example, each recipe has a title that is related to one of the chapters in which it falls: "Peaceful Vegetarian Lentil Soup," "Nostalgic Chicken Livers," "Diplomatic *Magloubeh*," "Poetic Baklava." Second, the recipes step outside the narrative to address the reader directly and elicit an engagement with the recipes, the described dishes, and the social contexts from which they emerge. The recipes emphasize the material process of cooking and, structurally, interrupt the growing-up and assimilation story of Diana.

Food is a trope, in the memoir, around which several affiliations are formed, severed, and crisscrossed along the lines of "origins" and racial/regional belongings. Anita Mannur, in "Culinary Nostalgia: Authenticity, Nationalism, and Diaspora," coins the term "culinary citizenship," which she defines as "a form of affective citizenship which grants subjects the ability to claim and inhabit certain subject positions via their relationship to food" (13). Borrowing Mannur's concept, Diana's grandmother and father stand for two senses of "culinary citizenship": the former insists on classification and categories, while the latter hints at multiplicity and complexity. "Gram is a baker, Bud's a cook." While bakers are "measured, careful, rational, precise," cooks are "dashing, improvisational, wayward, intuitive" (90). Diana's grandmother, though she knows—or perhaps does not know—that Muslims do not eat pork, prepares a big dish of "glistening, clove-studded ham" on Diana's father's first visit to their house (90). The traditional act of cooking acquires an "ethnic" connotation in this scene to reveal the cultural tension between the grandmother and Bud whom she perceives as "not precisely white." This, in a way, rehearses the movements by food reformers to unify a national cuisine for "American" citizens. Like the nineteenth-century food fights, which Gabaccia analyzes, where American values are asserted by means of following specific diets, the fights in Abu-Jaber's memoir (between Bud and grandmother; Bud and his neighbors) are also over food and the values each eating tradition stands for.

Borrowing from Roland Barthes's "Toward a Psychology of Contemporary Food Consumption" (1961), food in *The Language of Baklava* "has a constant tendency to transform itself into situation" (26). For example, in an incident in which Diana's father grills some food in the front yard, the neighbors are irritated by that behavior, which they perceive as *un-American*. Jaime, Diana's schoolmate and one of the neighbors' daughters, tells Diana in the school bus: "you better know that in this country nobody eats in the front yard. Really. Nobody. . . . If your family does not know how to behave, my parents will have to find out about getting you out of this neighborhood" (82). Throughout the memoir, Abu-Jaber gives numerous tips for being/becoming American. Thus, these tips are themselves national recipes for "American" way(s) of life.

Among the tips that Abu-Jaber includes to construct her "*American self*" (135) are consuming and observing closely "American culture, TV., music" (134). Especially in college, Diana has "finally acquired hip-hugger jeans and a long shag haircut, in the posthippie fallout look of the seventies" (135). Jeans, haircuts, clothing style transmitted through mass media and popular culture stand for what Stuart Hall calls "codes," which Diana emulates to construct her "*American self*." In "Encoding/decoding" (1980), Hall holds that ideology "once socialized, it becomes a code"; it is "a system of coding reality" (102). Despite the valorization of individual liberty, it is only by sharing the dominant "codes" of what *looks* American that Diana is deemed American or not quite. This unfolds a conundrum between the principle of individualism as foundational to the notion of Americanness and—what I have quoted earlier from Pease—"domestic Americanization."

Ironically, despite the actual diversity of American culture, this diversity is used to bifurcate American culture into "American" and "ethnic." For example, Diana's grandmother wants to educate her granddaughter about the "Orient." She decides to take Diana to downtown New York to watch the famous opera show "Madama [*sic*] Butterfly at the Metro-Politan" (93) and dine in a Chinese restaurant, the "Imperial Palace" (98). The choice of this particular opera in which the incidents of the story take place in Nagasaki of the early 1890s, when the Japanese navy claimed the city, implicitly recalls the victory of the United States in World War II after bombing Nagasaki in 1945. The circulation of this specific show in the Metropolitan Opera house in New York since 1907 can imply a commodification of diversity in cosmopolitan New York or a celebration of an imperialist America subduing the Japanese Empire. More importantly, the circulation of this opera perpetuates what Aki Uchida calls "the orientalization of Asian women in America" (1998, 161). In the story of Madame Butterfly, the Japanese wife is abandoned by her white husband who marries a white woman. Madame Butterfly gives up her child to be raised by her ex-husband and his new wife. The story recuperates the submissiveness of "Oriental" women as well as white superiority. Diana's grandmother's perception of the "Orient" and "Orientals" is also an example of orientalization. She praises "Orientals" as "*dainty* and *refined*," "[l]ike *little porcelain dolls* with their *little shoes and parasols*" (97 emphasis added). However, this simile between "Orientals" and "porcelain dolls" dehumanizes "Orientals" in the sense that they are compared to static artifacts that look alike. In other words, the term "Orientals" becomes a signifier for objects rather than people or for people reified as objects. In the "Imperial Palace," a Chinese restaurant, the grandmother tells the waiter that she has watched a show about "[*his*] people" and tells him how "spec-ta-cu-lar" and "ex-tra-va-gant" the show is (99 emphasis added). Making it a teaching moment, the grandmother tells Diana "[w]e feel it's more polite to call them Or-i-en-tals" without making a distinction between the Japanese and the Chinese (101). Drawing a distinction between her people and his people, as well as homogenizing "Orientals," is "racism with a distance," to borrow Zizek's words (44). The grandmother's remarks imply that "we" refers to individuals who identify as members of the nation-state (read: "American") and "them"/"his" refers to individuals who identify with their "particular" communities (read: "ethnic"). By the same token, the slow enunciation of "Orientals," "spectacular," and "extravagant" is a reminder of the first scene in the memoir where the broadcaster mispronounces Diana's last name. Both incidents exemplify the contradictions in American Orientalism: "deeply committed to U.S. primacy and to multiculturalism," in Vijay Prashad's words in "Orientalism" (2007, 176). In addition, unlike European Orientalism, as Uchida notes, "the Oriental geographically exists in the West" (161). American Orientalism takes shape inside the geographical territories of the United States rather than in colonies. Diana's and her grandmother's excursion echoes the American Orientalist discourse as it developed around the migration of the Chinese as indentured laborers.[10] The grandmother homogenizes "Asians" by assuming that the Chinese and the Japanese are one and the same people. Moreover, not only does she emphasize the exoticism, otherness, and foreignness of the "Orientals" but also reiterates the effeminate stereotype of "Oriental" men ("not like men at all," she says (104)). Homogenizing, stereotyping, and emasculating the "Other" are strategies that resonate with colonial national ideologies of American Orientalism that expanded in the nineteenth century as the United States emerged as a global power. Ironically, the grandmother tries to imbue her half-Jordanian granddaughter with this legacy of cultural imperialism.

The grandmother's intention to educate Diana about "Orientals" resembles a touristic excursion to an exotic place that simultaneously ethnicizes and empties the signifier "Orientals" of its meaning. Mieke Bal, in "Food, Form, and Visibility: *Glub* and the Aesthetics of Everyday Life" (2005), explains how "ethnic" restaurants are a "genre of 'tourism at home'" and a "form of neocolonialism." She argues that "ethnic" food is "a visible token of foreignness" that is not only perceived as enriching the diversity of the host country but also "incorporated [and] naturalized as a welcome element of otherness" (53). If "ethnicity" created by the host country is as Bal claims a form of neocolonialism, then Diana and her grandmother exemplify these dynamics between ethnic restaurants and their customers. Reading the menu, Diana reads "a treasure map that takes me [Diana] on its dotted line over snowy mountains, through hushed trees, past jade lakes" (98). Comparing the menu to a map materializes "Oriental" food in Diana's imagination. The topographic description of "Oriental" food in terms of mountains, trees, and lakes situates it as exotic food carried by people coming from faraway lands. Seeing and tasting the food arouse all of Diana's senses, "touch[ing] all the hidden places in [her] mouth" (102). At the center of the tourist experience is the "tourist gaze" as John Urry holds in *The Tourist Gaze* (2002, 1). One of its characteristics is that the "tourist gaze" is "constructed through signs" (3); hence, tourists search for specific venues and landscapes. The grandmother's *tour package*, visiting specific sites such as "Oriental" theaters and restaurants in metropolitan New York is allegedly to pursue knowledge about a "particular" community and to transmit this to her granddaughter. In fact, the grandmother is purchasing an "Oriental" cultural experience, which is not only pertinent to the tourist experience but also to a Eurocentric conception of "the Other as a self-enclosed 'authentic' community," to quote Zizek (44). In a complementary sense, the restaurant—Imperial Palace—markets itself as a touristic place by virtue of its name and its location. It is what Dean MacCannell calls "staged authenticity" (qtd in Urry 9). Urry elaborates MacCannell's concept and explains that "staged authenticity" aims at taking "advantage of the opportunities it presents for profitable investment" (9). Setting the stage for an authentic "Oriental" experience, the fortune cookies that Diana and her grandmother open are written in Chinese. Reading the slip—"no blame"—and translating it as "everything taste good if you hungry enough" (108) symbolically render the "ethnic" "Oriental" food generic and commercialized.

Nevertheless, there are two moments of resistance in this scene. First, Diana has a different consciousness of the waiter and the chef. Beginning to realize that the grandmother is talking about a Japanese, not a Chinese opera, Diana, the waiter, and the chef share a smile, as if they are "in on the joke together" (107). The subtle understanding between Diana, the chef, and the waiter in the scene renders the grandmother a member of a particular "ethnic" community (white ethnics). The second instance of resistance is when the chef and the waiter speak in Chinese, having an aside conversation among themselves, which only they can understand.

This experience of consuming "Oriental" food problematizes the "ethnic" as a qualifier for national identity. In the same sense, the "native" as foundational to the notion of "national" identity is also problematized in the memoir. "Native" suggests a point of origin and authenticity. By extension, it defines where a native should belong. Diana is born in the United States. By virtue of land of birth, Diana is a native of that land and therefore "American." However, Diana's father, as he sets the rules for his daughters, has taught them that they "are Arab at home and American in the streets" (5). To emphasize their Arabness, Bud does all the cooking and all the cooking follows Jordanian recipes. I interpret this, paradoxically, as a feminist gesture in the memoir that reverses gender roles and dismantles the arbitrary relation between women and domestic space. Yet, it can also mean more patriarchal hegemony over the domestic realm that allows Bud to control how he raises his daughters by imposing only Jordanian food on his family (which is all women: wife and three daughters). Bud's cooking, in addition, operates like "cultural mnemonics";[11] he cooks to remember Jordan. However, as I have pointed out earlier, Diana is not perceived, by her British friend, as a native of Jordan. The mere fact that the purpose of Bud's relocating the whole family in Jordan to familiarize his daughters with his own land of origin implies that his daughters are non-national in Jordan, *not yet* Jordanian. This is even underscored by Munira's description of Jordan in olfactory terms that Diana is unfamiliar with: "The original scent of Jordan is *here* [Jordan]: sesame, olive, incense, rosewater, orange blossom water, dust, jasmine, thyme" (37 emphasis added). *Here* implies that it is not only Diana who is alienated from the authentic scent of Jordan but also her father since he does not reside in Jordan. That is to say, both Diana and Bud acculturate to both Jordan and the United States as they move between the two nations.

Bud wants to fulfill the American Dream, as I have previously mentioned, by means of having his own restaurant. His dream restaurant, as Bud imagines it, "will be a real breakthrough, an amazing modern combination of Arabic and American food" (169). Bud's restaurant is not only a combination of Arabic and American food but also "a Shangri-la that finally heals the old wound between East and West. All languages will be spoken here, all religions honored" (172). This imagined restaurant evokes multiple connotations. It conjures up the prevalent notion of America as the land of opportunities that welcomes and hosts all religions, all languages, and everyone, whether from the old world or the new world. This implies that Bud's restaurant is a site for a heterogeneous imagined community. In addition, because it will be run by his family, it is also a congregation site for his daughters. That is to say, the restaurant will transform Bud's perception of America as a place of foreignness to a place of fulfilled dreams and belonging. Nevertheless, the fact that the restaurant is run by Bud's family duplicates the family dynamic—specifically his authority over his daughters—to reaffirm Bud's dominance. Moreover, the reference to "Shangri-la"[12] metonymically means not only a utopian place but also an exotic oriental land. Thus, simultaneously, the hospitality of the land of opportunities is intertwined with ethnicization.

While Bud dreams about a family restaurant in Syracuse, his own house in Jordan has turned into a restaurant—Kan Zaman. Visiting Jordan and the restaurant, Bud fantasizes about owning that restaurant, erected on the land of his ancestors. Yet, the restaurant's name and its tent-like entrance ethnicizes the place to attract more tourists. Echoing the Imperial Palace Chinese restaurant in Metropolitan New York, Kan Zaman is a Jordanian restaurant, located on the desert highway in Amman. Visiting his family in Jordan, Bud finds out that the house has turned into a restaurant—Kan Zaman [Once upon the Time]—in a touristic neighborhood in Amman and is commodified as an authentic ethnic place by virtue of its name and location. That is to say, within Jordan, the Jordanian restaurant becomes *ethnic* to sell an experience of authenticity to tourists in Jordan.

Diana's visit to Jordan as a writer in residence complicates the question of the national in the memoir even further through the example of the Sri Lankan maid who works for her uncle Jimmy (originally Jamil). Sri Lankan young women are "shipped" (256) to Jordan through a "slave agency" (260) to be adopted by Jordanian families for whom they work as maids. Many of these women leave their children in Sri Lanka while they work in Jordan. The uncle's Anglicized name—Jimmy—as well as the resonance between the living conditions of the Sri Lankan maids in Jordan and the history of slavery in the United States—specifically ill-treatment and separating slave mothers from their children—reproduce the binary between masters and slaves that deny the latter any sense of national belonging and rights.

The Sri Lankan maid cooking for uncle Jimmy's family adds another layer of complexity to the question of the non-national in Jordan. The question is complicated specifically in terms of the Sri Lankan cook's relationship to the food she cooks by virtue of which she can "claim and inhabit certain subject position," to quote from Mannur's theory of "culinary citizenship" (13). Unlike Diana's and her mother's cooking pancakes in Jordan that has facilitated the formation of new affiliations beyond geographical boundaries, cooking in Uncle Jimmy's house does not facilitate the formation of any communities around food. On the contrary, food in Uncle Jimmy's house rather evokes a sense of estrangement than affiliation. Diana and her friend, Audrey, are eager to leave, rather than stay to finish their dinner.

Jordan as a site for the national is complicated not only by the presence of immigrant labor but also by the widely spread American mass media—specifically soap operas and news networks—which unfolds a disjuncture between what Arjun Appadurai calls "mediascapes" and "ideoscapes."[13] Diana's uncles, when they convene, are always engaged in intricate discussions about nuclear weapons, oil crisis, and political turmoil; and wonder: "why is it . . . that America gets fatter, that American TV shows get louder, and that TV contestants win millions with a single answer, while rest of the world gets leaner, hungrier, sicker, angrier?" (274) This contradiction between the global political situation and the "chimerical, aesthetic, even fantastic objects," to borrow Appadurai's words (35), mirrors another disjuncture between the luxurious lifestyles portrayed in daytime American dramas such as *The Bold and the Beautiful*, which is very popular in the Arab region and the Middle East, and the lives of many people in Jordan, "a country crowded with Palestinian refugees [where] . . . so many people are hungry" (240). In Appadurai's words, "the lifestyles represented on both national and international TV and cinema completely overwhelm and undermine the rhetoric of national politics" (40). This disjuncture uncovers ways by which the national crisis of the Palestinian refugees in Jordan is subsumed by the fantasy of life in America that is conveyed in American mass media.

In conclusion, the non-national subject in *The Language of Baklava* is formed at particular moments of incompletion in its interpellation process(es); these moments are marked by tension between "ethnic" belonging and "national" identifications in both the United States and Jordan. This creates new sites for identifications that disrupt the growing up and assimilation narrative in more than one way. Structurally, *The Language of Baklava* does not open with a chapter about the protagonist's past life before coming to the United States or about a country of origin. Intertwined with this nontraditional opening is a blurring of the demarcations between the qualifiers "ethnic" and "national" by virtue of the formation of communities and affiliations transnationally, confined neither to the geographical boundaries of Jordan nor the United States.

An American Brat

Pakistani American author, Bapsi Sidhwa migrated to the United States in 1983. Sidhwa belongs to the Parsi minority in Pakistan, an economically and socially privileged minority whose roots are in Persia—current Iran—but have fled to the Indian subcontinent after the Arabs' conquest of Persia and imposing the Muslim rule. After the colonial partition of the subcontinent in 1947, Parsees lived diasporically between India and Pakistan, as well as in different countries around the world. Because of their very small numbers and ban on exogamy, they encounter the challenges of living as a minority, such as adjusting to the cultural traditions of the countries they live in, the threat of persecution, and political and social marginalization. Sidhwa has written four novels along these lines: *The Crow Eaters* (1980), *The Bride* (1983), *Cracking India* (originally title *Ice Candy Man*) (1991), and *An American Brat* (1993). Best known for her novel, *Cracking India*, Sidhwa in all her novels gives her readers an insider's story of Parsees' cultural tradition and Zoroastrian faith. Sidhwa's novels, except *An American Brat*, portray the traumatic events that have associated the communal strife and bloody partition of the Indian subcontinent and the creation of the new Muslim state of Pakistan. Sidhwa expands this postcolonial, South Asian literary tradition by narrating the partition story from the perspective of Parsee characters who have been separated from their family members in India and relocated in Pakistan. Unlike Sidhwa's first three novels, *An American Brat* does not highlight India's partition as its central theme. It rather focuses on the cultural differences between the United States and Pakistan, while shedding light on the lives of diasporic Parsee subjects in both states.

An American Brat revolves around Feroza's growing-up story that is interwoven with her going-to-America story. It portrays the protagonist's struggle with the Islamization of Pakistan and notions of choice and individualism in the United States. Though published in 1993, *An American Brat* has not received wide scholarly attention, unlike *Cracking India*, published earlier in 1991. Most reviews of *An American Brat* emphasize two major themes in Sidhwa's novel: the adjustments made by the protagonist to make herself fit in a new culture and the startling differences between the conservative East and the liberal West.[14] In most reviews, the novel is read as an assimilation narrative, where Feroza decides to stay in the United States and accepts living away from the rest of her family. However, as I show in my discussion here, the novel is not merely an assimilation narrative, but rather reveals Feroza as possessing a diasporic sense of imagined community that imposes on Feroza dual moments of estrangement and assimilation in both Pakistan and the United States and yields glimpses of a non-national consciousness.

Feroza's sense of imagined community is diasporic by virtue of being a Parsee:[15] members of her family are dispersed among Pakistan, India, and the United States. Nilufer Bharucha explains in "'When Old Tracks are Lost': Rohinton Mistry's Fiction as Diasporic Discourse" (1995b) that Parsees have formed three diasporas: first, the precolonial India diaspora (when they fled Iran); second, the division by partition of India; and third, Western/First World diaspora (in Britain, Canada, United States, Australia, and New Zealand) (57). One can therefore infer that Parsees, scattered as they are in various countries, belong to multiple communities concurrently within and outside the country where they reside, which in turn troubles monolithic identifications as solely Parsee, Pakistani, or Indian. Sidhwa, herself born in Lahore before the partition, defines her identity ethnically, nationally, and regionally; or as she refers to her identity, a "3P identity": "I am a Parsi first, then a Pakistani, specifically a Punjabi" (qtd in Randhir Singh 6). Many critical interpretations of Sidhwa's works—such as *Bapsi Sidhwa* (by Randhir Singh), *The Novels of Bapsi Sidhwa* (by Rajinder Dhawan and Novy Kapadia), and "Border Work, Border Trouble: Postcolonial Feminism and the Ayah in Bapsi Sidhwa's *Cracking India*" (Ambreen Hai)—trace a direct correlation between the writer's Parsee-Pakistan-Punjabi heritage and the storylines of her novels, which is not only clear in *An American Brat* but also in *Cracking India* (1991), *The Crow Eater* (1982), and *The Pakistani Bride* (1983). According to these critical reviews, Sidhwa's fiction familiarizes its readers with the history and culture of the Parsee diaspora in India, and in Pakistan, as well as in the United States, Although Parsees nationally belong to India and/or Pakistan by virtue of citizenship, right-wing Hinduization in India and Islamization in Pakistan have further minoritized Parsees in both countries. Feroza's dilemma of national identification in Pakistan occurs only after Pakistan became an Islamic state under General Zia-Ul-Haq. The Islamization of Pakistan has caused Feroza's feeling that she does not fit in Pakistan unless she adapts to the new Islamic rule and, in Zizek's terms, has created tension between her primary "particular ethnic identity" and her "universal identity as a member of a Nation-State." For example, Feroza wants to adopt a strict Islamic dress code imposed by the Pakistani government, whereas Zareen, her mother, disapproves of orthodox traditions. In an argument with Feroza about how to dress, Zareen contends, "we're Parsee, everybody knows we dress differently" (10). This moment of contention between Feroza, who dislikes the sleeveless blouse and sari, and her mother, who disapproves the fundamentalist turn in Pakistan, problematizes their affiliations with an Islamic state in two ways. First, dictating an Islamic dress in

Pakistan means the Islamization—or in other words, the desecularizing—of the "public sphere," to borrow Jürgen Habermas's term.[16] Second, the imposition of fundamentalist traditions on the "public sphere" has resulted in the exclusion of religious minorities such as Parsee, and of secular Pakistanis. Zareen, in order to save Feroza from the "puritanical," "mullah-ish mentality" (13) decides to send her daughter to the United States, which she sees as a more secular place. Despite this binarism between the Islamic state of Pakistan and secular liberal America, there is an interlocking relation between the United States and Pakistan, specifically the circulation of capitalism and fundamentalism.

The political changes in Pakistan that brought Zia-Ul-Haq to power and led to the execution of Zulfikar Ali Bhutto (the more liberal political leader in Pakistan) were supported by and took place under the auspices of the United States.[17] To combat the spread of communism in the region, Pakistan became the third biggest country—after Israel and Egypt—to receive US aid as well as a new market for the spread of commodities from the United States. The spread of commodities from the US and pro-American Pakistani government were accompanied by the Islamization of Pakistan, thereby "turning religious identities into political ones" as Mahmood Mamdani holds in *Good Muslim, Bad Muslim* (249). Politicizing Islam in Pakistan has led to minoritizing the Parsee community as well as jeopardizing secular persons who protested this regime. The circulation of Islamization in Pakistan is interrelated with the circulation of American capitalism. In the novel, there are references to "secondhand American garments" stores (13) and American and British videos (17). One can therefore infer that Feroza's encounter with capitalism and consumption starts in Pakistan rather than after her going to the United States. However, both ideologies—Islamic fundamentalism and capitalism—contribute to Feroza's feelings of alienation and estrangement within Pakistan and the United States.

Feroza feels a stranger and the need to adapt in both Pakistan and the United States. In Pakistan, Feroza feels a stranger, after the Islamization of Pakistan (because she is Parsee); and she feels a stranger in the United States (because as Manek, her uncle, says, she is “a Paki Third Worlder” (27)). Within the course of her journey between Pakistan and the United States, Feroza encounters moments of estrangement in the United States because she is a foreigner and within her Parsee community because of marrying a non-Parsee. These moments of estrangement, I argue, disrupt the assimilation story and complicate the formations of “imagined communities” to unfold instances of the non-national subject.

Upon her arrival at the John Kennedy airport, Feroza, not an American citizen, is non-national in the literal sense of the term. Passports, as Vijay Mishra explains in *Literature of the Indian Diaspora: Theorizing the Diasporic Imaginary* (2006), indicate “citizens whose bodies signify an unproblematic identity of selves with nations” (184). A noncitizen, Feroza in New York’s airport feels a stranger “in a strange country amidst strangers” (54). This collective feeling of anonymity, among strangers, is, paradoxically, at the core of Feroza’s sense of freedom. She feels free from the “gravitational pull” of “the thousand constraints that [have] governed her life” (52, 58). Feroza at this moment has an intertwined feeling of being both a stranger and a free individual.

This feeling of being both a stranger and a free individual recalls Gerog Simmel’s correlating freedom to being a stranger in his essay “The Stranger” (1950). For Simmel, the stranger is a figure emblematic of modernity in urban cities at the turn of the twentieth century. Feroza’s shopping expedition on Fifth Avenue resonates with the figure of the stranger in two senses: as a subject who is simultaneously attached to and distanced from life in the metropolis; and as a subject whose relationships are based on sharing common—rather than specific—qualities with other strangers.

Feroza’s ride with Manek into Manhattan is, for her, like “climbing into a futuristic spaghetti of curving and incredibly suspended roads, mile upon looping mile of wide highway that weaved in and out of the sky at all angles…and sometimes they appeared to be aiming at the sky” (67). Feroza’s perspective is that of the newcomer, who is fascinated by the vast and entangled layout of

New York city (spaghetti of roads and highways). Her description of New York positions Feroza as an observer who is *not yet* part of what she describes. In Simmel's words, Feroza "embodies that synthesis of nearness and distance which constitutes the formal position of the stranger" (404). This dual moment of proximity and distance is even clearer when Feroza feels both shocked and bewildered by the filth and poverty of Eighth Avenue and Forty Second Street, in comparison to the luxury and opulence she has seen in Fifth Avenue. It is "an alien filth" that unveils "the callous heart of the rich country" (81) to which Feroza cannot relate. It is a moment of (un)imagining communities that disrupts her assimilation and triggers her feeling of vulnerability. Feroza's vulnerability can be interpreted as a moment of fear caused by encountering the exploitative face of capitalism. That is to say, behind the enchantments of the United States (highways, skyscrapers, abundant food and merchandise, and efficient infrastructure) there is also a sense of confusion, disorientation, hollowness, and loss. Metaphorically, trapped in the stairways of The Young Men's Christian Association (YMCA), Feroza feels that "America assume[s] a ruthless, hollow, cylindrical shape without beginning or end, without sunlight, an unfathomable concrete tube inhabited by her fear" (90). This perception of the United States echoes, but with a contrasting twist, Feroza's earlier perception of New York as a *futuristic spaghetti* of roads and highways that aim at the sky. Besides, the metaphor of the locked exit door in the YMCA signifies Feroza's in-betweenness, both in and out of this new world and the new terrains she treads. This moment of mixed feelings of fascination, confusion, and fear is also a moment of nostalgia to the network of family members and acquaintances in Pakistan.

Feroza's feeling of estrangement is not only within the United States but also within her Parsee community. Toward the final chapters in the novel, Feroza's complex romantic relationship with David problematizes her belonging to the Parsee community since a Parsee woman is excommunicated if she marries a non-Parsee. Nevertheless, a Parsee man can marry outside his religion. Marrying a *non* is a dilemma for a Parsee woman. When Feroza's parents receive the letter about her plans to marry David, a big family meeting is held in which stories are shared about Parsee women who are denied appropriate funerals and social status by the Parsee community for marrying outside their community. In her attempt to stop Feroza's marriage to David, Zareen flies to Denver. Reiterating the rhetoric of individualism, Feroza ridicules her mother's concern with heritage and pedigree. She says: "If you [Zareen] go about talking of people's pedigree, the *Americans* will laugh at you" (277 emphasis added). It is not Feroza who abandons her religion and community but it is rather sexism that renders her a stranger to her community. Zareen starts questioning interfaith marriage in the Zoroastrian doctrine to conclude that the "mindless current of fundamentalism sweeping the world like a plague has spared no religion, not even their microscopic community of 120 thousand" (305–306). Islamic fundamentalism in Pakistan and sexism in the Zoroastrian religious tradition complicate Feroza's national belonging and religious affiliation.

Feroza and David, although they surpass East–West boundaries in their relationship, are unable to overcome the traditional perception of roots and heritage as intrinsic to their sense of belonging. Consequently, what starts as an intriguing romantic interest between David and Feroza turns into an impossible relationship because of insurmountable cultural differences. The theme of cultural differences frames the whole novel. For example, Manek describes Pakistanis as "Third World Pakis" with a "snow-white Englishman gora complex" (26). Besides, in an argument between Feroza and her mother about Feroza's relationship with David, Feroza asks her mother to think in new ways because "It's a *different* culture," but Zareen responds, "It's *not your culture*" (279 emphasis added). Cultural differences in *An American Brat* can be interpreted through the lens of American capitalism that commodifies *difference per se*. "[C]apital has fallen in love with difference," as Jonathan Rutherford reminds us in *Identity: Community, Culture, Difference* (1990, 11). The novel, in a sense, criticizes the capitalist's ideology of multiculturalism. As Zizek holds in "Multiculturalism, or, the Cultural Logic of

Multinational Capitalism" (1997), "the ideal form of ideology of this global capitalism is multiculturalism" (44). *An American Brat* depicts New York as an intertwined site of capitalism and multiculturalism. Feroza's impression of New York is of "a kaleidoscope of perceptions in which paintings, dinosaurs, American Indian artifacts, and Egyptian mummies mingled with hamburgers, pretzels, sapphire earrings, deodorants, and glamorous window displays" (76). Blurring the images of art and artifacts with food, clothing, and jewelry conjures up what Richard Sennett concludes in "The Public Realm" about diversity in New York and its planning. In New York there is "linear, sequential display of difference"; as he elaborates, New York is a mix of difference and indifference, races "who live segregated lives close together, and of social classes, who mix but do not socialize" (269). Along the same lines, Peter McLaren, in "White Terror and Oppositional Agency: Towards a Critical Multiculturalism" (1994), holds that despite claims of diversity, the ideology of multiculturalism either has its premises on assimilation or collapses into universalistic humanism that paradoxically reenforces Anglo-American norms and essentializes differences (48–52). *An American Brat* seems not to break from essentializing difference and valorizing the US ideals of diversity. Structurally, the novel is an amalgam of stories that revolve around *difference* situating Pakistan in contrast to the United States. It opens with Pakistan as a starting point and ends with the United States as a final destination; thus, it resonates with a conventional coming-to-America/assimilation narrative. Unlike *The Language of Baklava*, the first chapter in *An American Brat* "dramatically cuts between origin (the country of childhood) and final destination [U.S.]," to borrow Rosemary George's words (144). The opening chapter in *An American Brat* encapsulates the dilemma of the Parsee diaspora as well as the crisis of rising Islamic fundamentalism in Pakistan versus the more secular, liberal America.

Furthermore, *An American Brat* shares some aspects of popular fiction, specifically, the romantic plot of Feroza and David. The romantic plot in *An American Brat* follows a familiar narrative pattern of pop fiction: a simple plot in which the young woman from the less privileged lands of Pakistan comes to the United States, meets her dream husband, and challenges her parents and family traditions of marriage. Along these lines, Janice Radway in *Reading the Romance* (1984) holds that romances are all about a "man and woman meeting, the obstacles to their love, and their final happy ending" (199). In a similar sense, Walter Nash pinpoints, in *Language in Popular Fiction* (1990), the romantic story "is nothing if not predictable" (4). The predictability of Feroza's and David's falling in love is precisely what renders *An American Brat* a popular romantic fiction. The coming-to-America narrative is intertwined with the popular romance genre to reassert the ideology of individualism especially for the female protagonist, who challenges her community and its religious doctrines.

However, Feroza's and David's love story does not end in marriage and living happily ever after. Unlike the typical closures in romantic popular fiction, Feroza and David cannot surpass their cultural differences. In Woodrow Wilson's terms, Feroza has "*not yet*" become American; she and David foreground their ethnic roots. That is to say, the unconventional closure disrupts both the assimilation narrative and the popular romance story; and as a corollary, allegorizes a disrupted process of "transubstantiation," to recall Zizek: the shifting from the "particular ethnic identity" to the "universal identity as a member of a Nation-State."

Moreover, the love story between Feroza and David reverses what Mica Nava calls "the Orientalist critical gaze" (25). Nava explains that the Orientalist gaze is a "panoptic controlling male gaze," "in which the oriental woman is cast as object of sexual desire" (26). Contrary to the Orientalist "panoptic" gaze, Feroza objectifies the white male body to her gaze. Feroza's and David's first meeting is described as follows:

> At Feroza's timid knock on the garage door, David Press *revealed himself, wearing only* his ragged shorts and a pair of square, metal-framed glasses. The longish gold-streaked hair that swept his forehead and framed his handsome face appeared, if anything, to enhance the wild effect of his *gleaming nudity.* (245 emphasis added)

While reading Sidhwa's description, one cannot help but follow the slow rhythm of describing David's physique and his slow appearance—*revealing himself, longish gold-streaked hair, gleaming nudity*—from his garage door. This portrayal renders the *white* male's body exotic and erotic. To extend the reversed metaphor, unlike the Orientalist portrayals of the exotic women from the East as languid, lewd, and/or oppressed, Feroza is described as "impassive, imperious" (247). Feroza's haughty manner unsettles the East–West dichotomy that frames the whole novel.

Similarly, *An American Brat* subtly exoticizes the United States. The "America-returned" Pakistanis perceive the United States as an "*exotic culture*" (171). Echoing the pancake cooking scene in Jordan in *The Language of Baklava* that ethnicizes *American* food, the perception of the United States as an *exotic* place dismantles, what Zizek critiques to be, the US "privileged universal position," granted by its ideology of multiculturalism (44). Multiculturalism sets the United States in the superior position that claims appreciation of *diversity*, yet is founded on a demarcation between the "ethnic" and the *American*. This reconfiguration of the United States as an exotic culture can possibly renuance the definition of national consciousness to be in terms of diversity and multiple identifications, not in terms of assimilation and "domestic Americanization." Whereas on the surface level of meaning *An American Brat* reads as an assimilation narrative, on another level it disrupts this narrative. To recall Simmel, Feroza ultimately decides that she—like the stranger figure—finds comfort living in the United States away from family, because her feelings of dislocation, unbelonging, and anonymity are shared by many "newcomers" like herself (312). The last chapter in the novel sounds like Feroza's and Manek's testimony about living in the United States and its advantages, which focus on opulence, material goods, freedom, and technology. The whole chapter echoes and sums up the earlier chapters in which Manek underlines the stark differences between the United States and Pakistan. In this chapter, it is Feroza reiterating the same notions, and relieved that she is not in Pakistan, observing "grinding poverty and injustice" and "disturbing Hodood Ordinances" (312). In this final chapter, Feroza refrains from arguing with Manek. On the other hand, Aban, Manek's wife and a *newcomer* to the United States, plays the same role that Feroza plays earlier in the novel. The implication is that characters in the novel follow the same linear journey: move to the United States; are challenged by a new lifestyle; adapt to the new life in the United States; and finally are unable to give up living in the United States. The linearity of this narrative does not break from traditional migrant narratives that propagate Americanization as the final destination.

Intertwined with Feroza's journey to the United States is "a journey into the English language and into the 'ethnic' narrative of successful progress" as George points out (136). In the novel, Manek tells Feroza that she is lucky that her roommate—Jo—is a "real American" (148). Like Diana in *The Language of Baklava* who has to work on constructing her "*American* self" (135 emphasis added), Feroza works on *performing* "Americanness" by emulating Jo's informal way of talking, calling names, and eating canned and frozen food. In a similar way, Manek lies about being Christian to sell Bibles for Christian families. To make his sales' talk appealing, Manek uses words and phrases that Christian families would positively respond to: "How is little Jim (or Bill or Barbara) doing?" Have you started him on solids? . . . The Reverend told me Kevin is a mighty smart boy for his age" (202). The foregrounding of a singular accent, way of talking, and a dominant faith reconfirms the notion of a unified national community. Thus, "America" becomes "the imagined nation signified by . . . a monolingual tongue (English or rather, American English) and a determined assimilation of all differences into this national story," as George underscores (136).

Feroza ultimately decides that she finds comfort living in the United States, away from family, because her feelings of dislocation, un-belonging, and anonymity are shared by many "newcomers" like herself (312). That is to say, by the end of the novel, the community that Feroza imagines herself identifying with is a community of *strangers*. In that sense, *An American Brat* is similar to

The Language of Baklava; the imagined communities in both novels are neither "boundary oriented" nor necessarily "horizontal." The "imagined communities" in *An American Brat* are formed around a shared feeling of being "amidst strangers."

Feroza's acceptance of being a stranger in the United States because others are newcomers and strangers too is similar to the final scene in *The Language of Baklava* in which Bud starts his own fast food restaurant. The ending of both stories is paradoxical. On the one hand, these traditional closures offer the protagonists a place within the imagined community of the nation as Martin Japtok holds in another context (137).[18] On the other hand, within the course of both novels, by virtue of the fact that these are two-world novels, these texts unsettle the notions of a homogeneous national consciousness and renuance the formations of the imagined communities to be *trans*national and among *strangers*. In other words, whereas structurally, especially in the dénouement, both texts partake of a narrative of integration, the form of individualism that both stories depict is contingent and situational, and its *national* consciousness is not preset. By the end of both novels, Diana, Bud, and Feroza accept being *both* American and Jordanian, or *both* American and Pakistani, without necessarily resolving the tension between the *ethnic* and the *American*.

CHAPTER 2

Reimagining the US National Time in *West of the Jordan* and *The Last Generation*

In Chapter 1, I have argued that non-national consciousness is a specific moment in which the subject remains immutably foreign despite citizenship and acculturation. Along these lines I have used the term non-national not only to mean moving beyond geographical territories but also to mean using a non-nation-centered perspective to examine communities formed and defined by their ethnic, racial, and religious affiliations, located both within and outside of specific geographical territories. I have also reconfigured the notion of "imagined communities" as "imagined (transnational) communities." This perception of the nation in terms of *trans*national communities stands in contrast to the *paradigmatic story*[1] of consensus and homogeneity of the American nation. Within the context of nation formation, Anderson draws an analogy between the idea of the nation "conceived as a solid community moving steadily down (or up) history" and the "idea of a sociological organism moving calendrically through a homogenous empty time" (26). Nevertheless, the conception(s) of the nation from the perspectives of the colonized, the exiled, the displaced, and the migrant disrupt the genealogy of the nation as a steady movement in history as well as the perception of it as a solid community. I argue, here, in my analysis of *West of the Jordan* (by Laila Halaby) and *The Last Generation* (by Cherríe Moraga) that the perception of the United States by Palestinian migrants and by Chicanos/as is tied to their history of displacement from their native lands and the movement in time in both texts contends with the forward time movement that is characteristic of US national history. In *West of the Jordan* and *The Last Generation*, the present moment, what preceded it, and what comes after are not successive. In this chapter, I examine the nuances of what I call non-national time in *West of the Jordan* (by Halaby) and queer time in *The Last Generation* (by Moraga).

If, in Anderson's account of the formation of the nation, linear time is intrinsic to national consciousness, non-national consciousness produces a different kind of temporality, which I call "non-national" time. Non-national moments are specific instances, which are at odds with what Lisa Lowe calls, in *Immigrant Acts: On Asian American Cultural Politics* (1996), the "temporality of assimilation" (6). Lowe holds that the "temporality of assimilation" universalizes the experience of immigration in the sense that different migrant groups assimilate in the same way; hence, racialization and hierarchies within minorities are obscured. In *West of the Jordan*, the characters' movement in time is dispersed between the West Bank—the occupied territories in Israel—and the United States. Hence, it does not have an ordered pattern; it is rather interstitial, fragmented, or cyclical. I root non-national temporality in Ernst Bloch's theory of "contemporaneous non-contemporaneity,"[2] which Harry Harootunian elaborates in "Some Thoughts on Comparability and the Space-Time Problem" (2005). Harootunian holds that "contemporaneous non-contemporaneity" are moments of different but coexisting temporalities, "where fragments of the past unexpectedly and suddenly rise up to impinge upon the present" (42). The 1967 setback and its aftermath—loss of Palestinian territories, and expansion of Israeli occupation—looms over *West of the Jordan* and is the core cause for multiple conflicts between migrant parents who demand that their children should acknowledge their Palestinian heritage and their children who are born and raised in the United States and perceive themselves as mainly Americans. Along the same lines, queer motherhood in *The Last Generation* produces a different sense of lineage formed at the conjuncture of imperial modern history of the United States as well as pre-Columbus and precolonial Aztlán. I situate non-national time and queer time within accounts by E.J. Hobsbawm, Donald Pease, and Homi Bhabha that criticize, and replace totalizing national time in Anderson's account with multiple temporalities.

Hobsbawm, in *Nations and Nationalism Since 1780: Programme, Myth, Reality* (1990), holds that "the progress of national consciousness…is neither linear nor necessarily at the expense of other elements of social consciousness" (130). That is to say, within national time there are multiple temporalities formed by virtue of multiple aspects of "social consciousness" such as occupation and displacement. In *West of the Jordan* and *The Last Generation*, the story of the nation is expanded beyond confining geographic territories, hence dispersing the notion of national time between two different time zones (as in *West of the Jordan*), and it criticizes American imperialism (as in *The Last Generation*). Non-national time and queer time are two specific moments in the nation's time that are "interruptive" and "disjointed" to borrow Pease's words (154). Hence, national time would be imagined not as a single but as a "double" time: pedagogical and performative, as Bhabha holds in *The Location of Culture*. On the one hand, in pedagogical time, "the people are the historical 'objects'…giving the discourse an authority that is based on the pre-given or constituted historical origin *in the past*" (208). On the other hand, the performative time "introduces a temporality of the 'in-between.'…[that] interrupts the self-generating time of national production and disrupts the signification of people as homogeneous" (212). *West of the Jordan* and *The Last Generation* are narratives that are written from the perspectives of the displaced, the colonized, and the migrants: Palestinians and Chicanos. Accounting for these perspectives reshapes the US national time to become a temporality intertwined either with Jordan and Palestine or with Latin America and the Caribbean.

Despite their different histories in the United States, both Laila Halaby and Cherríe Moraga share the experience of displacement and dispersal because of the Israeli occupation of Palestinian territories or US imperial colonization of one-third of Mexican lands. Loss of land and sovereignty over their own territories is at the heart of the Palestinian experience of exile, diaspora, and migration. When Israel established itself as a Jewish state in 1948 and then expanded its occupation of Palestinian territories after the Six-Day War in 1967, Palestinians fled to several neighboring countries that gave them citizenship in the country of residence—but eventually deprived them of that citizenship—or gave them work permits with various restrictions on the kinds of jobs they were allowed to take. In a similar sense, the neo-imperial project of manifest destiny seized Mexican lands by virtue of the 1848 Hidalgo Treaty. What were once Mexican territories became American territories, and members of the same family residing on Mexican lands became residents of American territories, while other members of the same family who became residents of Mexican territories became foreigners to what were once their own lands. Both writers take as a central theme the passage of heritage from parents—specifically the mother figure—to children to reaffirm belonging to a specific land and its cultural heritage. However, this passage of heritage from one generation to the next is neither homogeneous nor linear. On the contrary, the relation to cultural heritage and to parents as bearers of that heritage has a complex pattern because its genealogy does not move "calendrically" as Anderson holds but rather *recursively*, as in *West of the Jordan* or *monumentally*, as in *The Last Generation*.

National time in *West of the Jordan* and *The Last Generation* goes beyond the geographical boundaries of the United States and unfolds in the historical and cultural interaction between the United States and either Jordan and the West Bank (as in *West of the Jordan*) or Latin America and the Caribbean (as in *The Last Generation*). In *West of the Jordan*, the frame story of Hala migrating from Jordan to the United States alludes to the 1967 setback that has brought further displacements for Palestinians. The novel is about four Palestinian cousins—Hala, Khadija, Soraya, and Mawal—whose extended family lives in diaspora in the United

States and Jordan. The novel covers their lives in three countries: the United States, Jordan, and Palestinian territories under Israeli occupation. Each family has relatives in each of these three countries with whom they exchange stories about their daily lives and experiences to maintain their social network and family ties both within and across these countries. Thus, their lives, stories, and histories are dispersed across the three geographical spaces. *The Last Generation* is a collection of poetry and prose addressing the indigenous heritage of Chicanos/as. It focuses on the usurpation of Chicanos/as' native lands five centuries ago (first by the Spanish, then by the Americans) and the deteriorating geopolitical conditions on multiple levels in the 1990s. There are two significant moments that Moraga continuously refers to throughout the book. These are the colonization of the Indian lands that took place in the sixteenth century and the political and social events that took place across the world at the beginning of the 1990s—the Soviet Union dissolved, the Cold War ended, the Gulf War started. Time is shaped by conquest, invasion, occupation, and political unrest in both the occupied territories of Palestine (in *West of the Jordan*) and indigenous Indian lands (in *The Last Generation*). To examine non-national time and queer time in *West of the Jordan* and *The Last Generation*, I read the former as a moment formed at the intersection between an ongoing occupation of Palestinian territories and continued adaptation to living in the United States; and I read the latter as a moment formed at the conjuncture between the precolonial and the colonial.

West of the Jordan

West of the Jordan, published in 2003, caught the attention of reviewers as a novel by a writer with Arab roots that contributed to the emerging literature by Arab-American women writing about Arab culture and heritage to challenge the stereotype of women from the Arab world. This is Halaby's first novel, and her self-described purpose for writing is to challenge "the perception, or misconception, of Arab women vs. the reality of Arab women" (online interview). She also aims at portraying Arab families closely and exploring the effects of occupation and exile on the Palestinian family relationships. The novel has been received very positively and most of the reviews[3] highlight its main themes, which can be summed up as follows: the difficulties of cross-cultural existence, multiplicity of Palestinian women's experiences, diversity of Palestinian women, and mother-daughter relationships.

Despite the positive reception of the novel, all these reviews, except for one by Fay A. Chadwell, do not acknowledge that two of the main characters are born and raised in the United States and hence struggle with defining themselves as solely "Arab." By the same token, none of the reviews discusses the historical/political context of the Israeli occupation of the West Bank or the interaction between characters living in the United States and their relatives living in Jordan and the West Bank. For example, Parama Sarkar's review discusses how Halaby's novel "reflect[s] the multiplicity of Palestinian experiences [and women]" (263). In addition, Sarkar describes Mawal as "immobile and located in a specific cultural space" (265). While the first comment ignores the fact that at least two of the main characters—Khadija and Soraya—are Palestinian/Jordanian-American (Khadija does not even know Arabic), the second comment misses how Mawal draws an analogy and a metaphorical proximity between the United States and Nawara, which brings into question the "specificity" of the cultural space that Sarkar refers to. In similar ways, Dori DeSpain writes that the main theme of the novel is "the difficulties facing *Arab women wherever they live*, . . . seeking acceptance and success in a *foreign country*" (208 emphasis added). According to DeSpain's reading of the novel then, Khadija who is born in the United States and does not know Arabic, is an "Arab" woman who lives in a "foreign country." These reviews unfold a problem in the critical reception of literature written by women with Arab roots. Amal Amireh and Lisa Suhair Majaj, in *Going Global: The Transnational Reception of Third World Women Writers* (2000), point out that texts by Third World women are viewed either as markers of "cultural authenticity" or as texts that provide "windows" into other cultures (2). The reviews of *West of the Jordan* contain clichés concerning texts by and about women of Arab roots. As Sarkar's and DeSpain's reviews reveal, they reaffirm a binary between the East that is "immobile" and the West that is a more progressive place for Arab women who are independent and ambitious. This *prepackaged* reading—as Mohja Kahf holds in "Packaging 'Huda': Sha'rawi's Memoirs in the United States Reception Environment" (148–149)—does not attend to the intricacies and nuanced lives of Palestinian-American women's writings.

In *West of the Jordan*, as previously mentioned, members of the same family are dispersed between the West Bank, Jordan, and the United States. Despite that dispersal, what brings members of these families together are watching wedding videotapes of their relatives in Nawara/Jordan and the circulation of gossipy stories. This suggests that despite the geographical distance between these three countries, that distance shrinks by virtue of exchanging videotapes and gossip. These dynamics too conjure up "imagined communities." As Anderson notes, reading novels and the newspaper bring people of the same community together; "imagined communities" form the core principle of the nation and what distinguishes communities is the "style" by which they are imagined (Anderson 6). The time difference between the United States and Jordan/West Bank shapes in a paradoxical manner the ways communities are imagined in *West of the Jordan*. On the one hand, the time difference between different time zones is overcome by technological means of communication. Thus, watching a videotape or circulating gossip is no longer obstructed by the geographical distance between the United States, Jordan, and the West Bank; a videotape can be (e)mailed overnight and gossip can take place over a phone call. These transnational transactions, made easier by contemporary technologies, challenge the constraints of spatial distance; as Paul Jay puts it: they "collapse the discontinuity between time and space in a radically new way" (36). Herein lies what I argue to be a non-national time—and paradoxical time that is interstitial, fragmented and nonsequential, and recursive and nonsynchronous—and through its lens I analyze *West of the Jordan*.

The preface of *West of the Jordan* foreshadows its peculiar temporality. Titled "Make It Delicious" rather than *Preface* or *Foreward*, this chapter is an inner dialogue in which the writer blames herself for not writing as frequently as she should, and she encourages herself to write stories inspired by her own experience of recollecting stories from past incidents and events. The structural purpose of this chapter is twofold. First, by fictionalizing the preface, the conventional opening of the story is disrupted because the preface becomes part of the narrative but not quite. Consequently, no linear progression of the preface can be traced in the following chapter. Second, the present moment that the author perceives as a moment of recollecting stories and writing them is depicted in contradictory terms. She writes:

> *now* they [stories] come back to you at the *wrong time*, at the *right time*, at times that make you *hate* where you live, or *love* it more than you can make your words describe. (1–2 emphasis added)

The contradictions of the present moment reveal that the "now" moment for the author is a site for *different but coexisting temporalities*. By the same token, the pleasure of recollecting and writing stories is compared to cooking a delicious meal. This analogy between cooking and writing conjures up the multiple communities that can possibly be formed around food and stories. These communities are formed simultaneously but not synchronically. In other words, the incidents in the novel are set in Jordan, the West Bank, and the United States. By virtue of their geographical locations, they have different time zones. When it is daytime in the United States, it is nighttime in Jordan/West Bank. There is little reference to specific clock-time—whether it is day or night—either in the United States or in Jordan/West Bank. For example, the incidents of watching a wedding videotape take place in Jordan, the West Bank, and the United States. However, the time frame of watching this tape is never given. While the act of watching can be simultaneous, it cannot be in the same clocked time. Consequently, the movement of the imagined communities formed around watching videotapes is not *steady* and is not formed in *homogeneous time.*

Time in *West of the Jordan* moves in circuits—incidents, stories, and acts—one leads to another and circulates among family members. There is a glimpse of this temporal manner in the first chapter, specifically, when Hala recalls the political and family incidents that have taken place in 1967 and have forever affected her family:

> In 1967, the Occupation began and my father lost a lot of his land. My mother lost her freedom to visit her family [in the West Bank]. My father became less generous with my mother, and she became less generous with her children. (11)

Furthermore, Hala's mother became terminally ill with cancer. The year 1967 has caused a temporal disruption in Hala's family as well as in the formation of the Palestinian nation. On the one hand, in Hala's family, roles have been reversed and her older sister, Latifa, has taken over their mother's role as a caregiver for both her siblings and their frail mother. On the other hand, since the Six-Day War, Israel has expanded its control over Palestinian territories and seized Palestinians' sovereignty to their own lands. The structure of the novel echoes this temporal disruption: the sequential order of chapters is disrupted and is either reversed or does not follow a specific pattern. The first chapter in the novel, which is narrated by Hala, is followed by a chapter narrated by Mawal, her cousin in Nawara, then Soraya, then Khadija. This sequence is repeated but in different order: Hala, Mawal, Soraya, Mawal, Khadija; Hala, Hala, Soraya, Mawal, Khadija, Soraya, Khadija, Soraya, etc. The order of incidents in the novel does not follow a cause–effect logic. They are stretched along the course of the novel and within the bigger frame of Hala's visit to Jordan that starts in the first chapter and her return to the United States in the last chapter.

Stretching one incident in the novel over a number of chapters fragments and expands the time it takes to narrate the incident. This temporality renders *West of the Jordan* a "spiral emplotment."[4]

A spiral narrative is a paradoxical one that has a sense of both linearity and circularity. The linear parts in Halaby's book, specifically Hala's visit to Jordan and meeting with different members of her family, are loaded with incidents or memories of incidents that took place in the past as well as reflections on them. This results in a turn away from the moment narrated and allows these recollections to interrupt the narrative. As Paul Ricoeur explains in *Time and Narrative*, "excursions into the past" like flashbacks and memories have two effects on the narrative. They "form a *series of loops* that gives its specific distension to the narrated time's extension" and "*paradoxically*, make the narrated time advance by delaying it" (103–104 emphasis added). That is to say, the present moment expands, intersects, and moves liminally between various stories. This is clear when reading together the first chapter of Hala's story and the last chapter of the novel, which is also about Hala.

The novel starts and ends with the same character: Hala. As I noted earlier, the story opens with Hala about to land in the Jordanian airport, in an in-between space between two destinations. The rest of the chapter is about her childhood memories, remembering her mother and her stories. Hala also remembers the devastating family and political incidents she witnessed as a child and feels the pain of losing her mother anew. The last chapter is also about Hala, her adjustment to a new life in the United States, and she recalls what her mother told her about the importance of remembering to make her days "new" and "old." The first chapter is titled "Going Home"—in that chapter, it is Jordan. The final chapter, where the novel reaches a dénouement, "home" is the United States to which she returns after her visit to Jordan. In this final chapter, Hala compares her uncle's—Hamdi's—upper-class house and its bare walls to her house in Amman, with "every nook and cranny filled with something: a plant, a book, a statue, a flower" (217). In an earlier chapter titled "Weddings," Hala perceives Jordan as a place "Of Unfulfilled Dreams" and the United States as a place of "Reason and Capitalism" (83). The tension in the novel stems from a tension between the "old" and the "new," allegorized by a spatial contrast between Hala's house in Jordan and her uncle's house in the United States. This spatial contrast implies two senses of temporality: linear and patterned (allegorized by Uncle Hamdi's house) on the one hand; emotional and instant (allegorized by Hala's house in Amman), on the other hand. Hala thinks of Uncle Hamdi's house as one that suits his "professor lifestyle," "high-class American style," neat, lush carpeting, and regulated temperature (216). This functional luxury is tied to affluence and generates a kind of temporality that is predictable and follows a set pattern. Despite the comfortable life that Hala has in the United States, she is nostalgic for the emotional connections she has in Jordan. In her uncle's house, despite the luxury of comfort and predictability, time and the movement in time are locked into a present that is empty. When Hala looks at the walls in Uncle Hamdi's house, there are no photos or colors and she "cannot imagine anything" (217). In contrast to that empty present, in Amman,

> every wall was covered with religious plaques, calendars, photographs. Every gift and every souvenir ever received or bought was on display like a trophy. Always somewhere to look to take you somewhere else, to make you think. Either a memory resurrected or a new place to go or a joy to feel.... No place for thoughts to stop. (217)

The circulation of affect—joy and pride—evoked by gifts and souvenirs conjures up what Arjun Appadurai in *The Social Life of Things: Commodities in Cultural Perspective* terms the "social potential" of commodities (6). Appadurai proposes defining a commodity "as the situation in which its exchangeability (past, present, or future) for some other thing is its socially relevant feature" (13). In that sense, Hala exchanges her feelings of nostalgia with her recollections of the associative meanings of gifts and souvenirs on display in her house in Amman. By the same token, the cluttered walls and memories reflect a structure of time that is cumulative and evocative compared to the bare walls and the neat surroundings that reflect an ordered structure of time. This is precisely a moment in which "fragments of the past unexpectedly and suddenly rise up to impinge upon the present," producing *different but coexisting temporalities*, as Bloch reminds us. Memories of Hala's house in Amman *impinge* on the present moment and extend the present moment into the past and paradoxically into a new undescribed future: "to take you somewhere else . . . a memory resurrected or a new place to go or a joy to feel." This moment of conjunction representing the past and the future recurs throughout the novel to reveal an aspect of noncontemporaneity between Nawara, which is under occupation and without an independent economy, and Glendale/California, which provides affluence and a seeming economic security.

Nawara—like all villages in the West Bank—is famous for stitching rozas. However, working at embroidery declines "for lack of *time* and *money*" (10 emphasis added). Patterns of embroidering rozas can be linear, cyclical, intersecting, or crisscrossing. These metaphorical multiple movements in time stand in contrast to the temporal matrix under capitalism that "seeks to unify and homogenize . . . and makes the movement of time a one-way street, all directed toward the product," to quote from Harootunian (45). Rozas metaphorically stand for the "spirit of Nawara" (15) and are set in contrast to Glendale in two senses: a politically occupied territory and a recipient of American money sent by the Palestinian diaspora in the United States, as well as the homeland of Palestinian cultural heritage. In this regard, Mawal gives a complex description of Nawara in terms of its geopolitical location and its rozas:

> The complicated embroidery on our *rozas—with both Palestinian and western stitches and patterns*—captures the spirit of Nawara, which sits at the top of the West Bank, just west of the Jordan River, east of Jenin and far enough away from both of these places to be a peaceful village that only ever so often releases an avalanche of stones and fire. This is something that happens more often as the Israelis take parts of our village to build their settlements. (15 emphasis added)

The art of embroidery in Nawara metaphorically reveals the constraints of both the political and economic occupation of Nawara.

This occupation complicates the formation of an "imagined community" for the Palestinian diaspora or its *steady movement down or up history.* Jewish settlements are perpetually disrupting the possibility of forming an "imagined community" within single geographical territories. What is formed is rather an "imagined [dispersed] community." As metaphors, rozas with their Palestinian and Western patterns unfold the multiple dispersals that are characteristic of the perception of Palestine as a nation without sovereignty.

Despite the economic and political occupation, new imagined communities are formed around watching videotapes, sent by their migrant relatives, and around the everyday stories and gossip shared among families in Nawara. Mawal says:

> I tuck this story [the story of installing speakers to a mosque's minaret] into my pocket, wishing I could stitch it into my skin, like one of the *Bedouin tattoos* my grandmother wears. Are there stories like this in lovely, tempting America? Do my cousins there even know these little histories? I doubt it.
>
> Stitch in red for life.
> Stitch in green to remember.
> Stitch, stitch to never forget. (103 emphasis added)

Bedouin tattoos, like embroidery on rozas, are intricate. They are encoded texts on one's skin that imply meanings of belonging, acknowledging, valorizing, and remembering a specific cultural heritage. Thus, Bedouin tattoos are the means to keep alive the "spirit of Nawara" and, by extension, to keep remembering people and their lives in Nawara despite the loss of land and the decline of embroidering rozas. Comparing stories to stitching and tattoos implies that these stories are imprinted on the skin of its bearers, removing them is more painful than keeping them. Thus, stories and skin become impossible to separate. The circulation of gossip as stories in the novel functions in the same way that Jörg Bergmann suggests in *Discreet Indiscretions: The Social Organization of Gossip.* Building on anthropological studies about gossip, Bergmann explains that gossip "strengthens the identity and coherence of a social group" (145). Nevertheless, as Bergmann notes, gossip has been also tied to premodern societies (141). Along the same lines, Gerald Arbuckle in *Violence, Society, and the Church* holds that gossip "flourishes in pre-modern cultures as an informal method of maintaining loyalty and conformity to the group" (99). In *West of the Jordan*, phones become part of the infrastructure of maintaining community by facilitating the circulation of gossip between Nawara and Glendale. Ironically, the gossip about Hala's mother having an affair with a man from her village started in the United States, hence disturbing the binary between the United States and Nawara.

By the same token, in Nawara, new communities are formed around watching videotapes, which can be interpreted in light of the technological development in the fields of transportation and communication, which "foreshortens time"[5] as Hans Meyerhoff explains (109). The paradoxical relation between "foreshortening" time and the expansion of time by remembering stories and incidents that have taken place in the past is tied to capitalist economic structures and new means of visual consumption. Nawara, as I have pointed out earlier, is famous for stitching rozas, a craft that declines "for lack of time and money" (10). Yet, every house, regardless of its economic status, has a TV and a VCR. These are the means of seeing "faraway sons, cousins, brothers, sisters, neighbors" (18). That is to say, what brings people together in Nawara is not the traditional activity of embroidery (which takes quite some time to finish) but rather another activity that has a shorter time span (watching videos) that has become more and more intensified in different ways due to the widespread capitalist economic policies that "foreshorten" time and necessitate fast transfer of money and information. Watching wedding videos and weaving rozas form "imagined communities" that consolidate the ones formed in Glendale/Hollywood/Anaheim, which Soraya describes and I will discuss later. This disrupts the assumption that members of the family who do not migrate are more settled, or feel more at home than their relatives who have migrated.

The simultaneity of different forms of temporality within a single space opens up the possibility of their comparability to unfold new movements in time. Within the context of the novel, there are three kinds of circulations: of gossipy stories, which is two way between the United States and Nawara; of money, which is unidirectional from the United States to Nawara; and of gifts, which is two way between Nawara and the United States. These movements in time, though they take place in discrepant time zones, become simultaneous, but not synchronic, by virtue of the technology that facilitates long-distance communications. Simultaneity and nonsynchronicity in the novel constitute a temporal matrix that dismantles the definition of the nation in terms of unidirectional temporality. To clarify, for example, many marriages as well as many houses in Nawara are built on American money (219, 22). Family members who have left for the United States send money to relatives left behind in Nawara to show that they have a better standard of living than other families who do not have the same advantage. The money sent by relatives in the United States intertwines the lives of individuals who live in the United States with those who live in Nawara.

Money and gifts sent to Nawara as a means of financial support of relatives left behind is also the means by which relatives who left for the United States show their connectedness to their land of origin. In this sense, the sender/United States is the source of economic support for the receiver/Nawara, while Nawara and its people are dependent on the United States for economic survival. Nevertheless, the videotapes/pictures sent from Nawara to the United States as well as the stories told to and by Mawal and exchanged with her cousins in the United States reverse this sender–receiver dichotomy because relatives in the United States are dependent on these videotapes and stories for their cultural and emotional survival. Throughout the novel, though Mawal acknowledges gifts and money sent from relatives in the United States, she emphasizes the superiority of Nawara over "lovely, tempting America" because the latter lacks stories about people's daily lives (103). Thus, Nawara becomes the *sender* and the United

States becomes the *receiver.* Monetary remittance (money and gifts) sent to Nawara as well as "social remittance"[6] (videotapes/pictures and stories) sent to Glendale/California become a "two-way" remittance that decentralizes the notion of a nation of origin and consequently the center-margin dichotomy intrinsic to the meaning of remittance. Kezia Page, in *Transnational Negotiations in Caribbean Diasporic Literature: Remitting the Text* (2011), draws on the term remittance[7] from economic discourse and uses the term "cultural remittance" to denote a two-way exchange of remittance—money remittance and cultural remittance—between the host country and home country (11). Not only does this "two-way" remittance decentralize notions of "national" home but also reflects the double uprootedeness of the Palestinian diaspora. Within Israel, Palestinians cannot easily move from one city to another, checkpoints have to be passed and sometimes a special permit is required. Thus, family networks are formed within a perpetual diasporic flow of people across borders whether the borders are within Israel or outside of it. In a similar sense, if one considers Halaby's novel as a "remittance text,"[8] consumed in Nawara, the homeland of the characters, *West of the Jordan* reverses that order as well. This is because, contrary to what Page holds about diaspora narratives as "cultural remittance," sent back to be read in the native lands of their writers, the target readers of *West of the Jordan* would be mostly Palestinian-Americans or Palestinians who are well acquainted with the English language. Thus, it becomes a "remittance text" consumed in what is considered home; in that context, it is the United States rather than Nawara.

Furthermore, *West of the Jordan* portrays an imagined proximity and similarity between Nawara and Glendale from the perspectives of Mawal and Soraya. Soraya, an American-born child, comments: "My parents are from the same village in the West Bank, and half of the village lives here in Glendale or Hollywood or Anaheim" (31). In another chapter, she says: "Our house has an endless supply of visitors, *as though this were Nawara*" (106 emphasis added). "[A]s though this were Nawara" evokes the presence of a Nawara imagined community, in Anderson's sense,[9] by virtue of which they affiliate themselves emotionally and mentally to a land of origin. However, this sense of an imagined community becomes problematic if one takes into consideration the fact that the West Bank,[10] where Nawara is imaginatively located and to which the title of the novel alludes, has a long, complex, political history of defining and claiming its sovereignty. On the one hand, the West Bank is a disputed territory between Palestine and Israel. On the other hand, Palestine is not politically fully acknowledged as a state; there is no Palestine on the world map and no Palestinian citizenship in the conventional sense. Thus, the novel depicts a metaphoric national community that only emerges within the narrative. This metaphoric national community is also formed outside its geographical territories, namely in California, in the United States

From Nawara, Mawal draws a metaphoric proximity between Nawara and the United States. In the chapter titled "Nawara," Mawal says: "Nawara could have a small version of herself in the United States, which is like an army calling all able-bodied young men away and then never returning the bodies" (15). Joining the army always implies combat and the possibility of facing death. Therefore, pairing the United States and Nawara in terms of joining the army is twofold. First, it implies their destinies are intertwined; thus, the safety of one place hinges on the safety of the other. Second, it shatters the presumed binary opposition between the United States as the land of dreams and opportunities versus Nawara as a small village in a war-torn zone. Furthermore, this comparison echoes what Soraya says about functioning of her parents' house as though it "were Nawara." Mawal's perception of the United States reveals a contradiction. On the one hand, the United States, though a haven for the Palestinian diaspora, does not lessen their feelings of alienation and dispersal. On the other hand, in a different chapter, titled "America," Mawal perceives the United States as "a *greedy neighbor* who takes the best out of you and leaves you feeling empty" (96 emphasis added). This metaphor reveals the feeling of emptiness—rather than fulfillment—that the United States, as an alternative country to live in, brings to the

Palestinian diaspora, and echoes Hala's perception of Jordan as a place of "unfulfilled dreams."

Not only does *West of the Jordan* cast doubt on the unidirectional movement in national time, but also complicates the generational transmission of the nation from mothers to children, specifically to daughters. In immigrant narratives, the mother figure is often emblematic of national and cultural heritage; hence mother-daughter conflict has the potential to interrupt national time. A case in point is the contrast between Khadija and Mawal in terms of their relation to their mothers. Khadija is perceived as a girl with "an Arab face" but with "an American accent" (19). Khadija's Arab features make her look familiar in Nawara. However, her fluency in English rather than Arabic disturbs that sense of familiarity. Khadija's character complicates the definition of Americanness when she paradoxically perceives Americanness in a homogeneous sense, yet defines Americanness in terms of diversity. At school, Mr. Napolitan

> expects me [Khadija] to know more than other kids because my parents *are not American*, though there are lots of other kids in the class who *aren't American* themselves. I want to scream at him that *I am just as American* as anyone here. (74 emphasis added)

Khadija's self-perception as an American problematizes the notion of Americanness. On the one hand, Khadija is American by virtue of being born in the United States though her parents are not American. On the other hand, this is what Americanness is: belonging to parents who come from any part of the world to the United States and hence *become* "American." At home, Khadija's mother tells her "You are Palestinian," and she replies "I am American." Khadija elaborates that she does not even know Arabic and that she is "American." Though Khadija tries to convince her mother that she can be "American" and still be her mother's daughter, her mother firmly responds: "No! No daughter of mine is American" (74). Khadija's mother rejects her daughter labeling herself "American" because being American implicitly severs cultural ties with her mother's Palestinian heritage. Unlike Khadija, Mawal sustains a close relationship with her mother; they share stories about their family members in Nawara, Jordan, and the United States; and their acquaintances in Nawara perceive them as sisters rather than mother and daughter. If one considers the meanings of Mawal's name (a folkloric ballad that is popular and broadly circulated)[11] and relate them to the metaphoric description of herself as a tree whose branches are breaking, one can conclude that her presence as a safekeeper of stories of her relatives and friends in Nawara/Jordan/the United States is essential to keeping the Palestinian heritage within and outside of Nawara.

Mother-daughter relationships in Halaby's novel do not follow the patriarchal family paradigm, in the sense that the novel is narrated from the perspectives of the four daughters—Hala, Khadija, Mawal, and Soraya—and not from the perspectives of the fathers, brothers, or husbands. In that sense, one can consider Halaby's novel a "feminist family romance,"[12] which unfolds different temporalities, formed along the lines of controlling women's sexualities and the role of the mother as a transmitter of national belonging. The conflict between Khadija and her mother reveals Khadija's rejection of bearing the Palestinian cultural tradition of her mother, not only because Khadija perceives herself as American, but also because she resists the patriarchal control over her sexuality. Khadija's father, when he sees her belly-dancing, feels the threat of his daughter's evolving sexuality and beats her. Through the act of violence the father imposes the phallic order that contains and represses female sexuality and relegates it to the realm of the invisible. Khadija's mother reemphasizes and even complicates that phallic order when she correlates the threat of her daughter's evolving sexuality to the threat of losing contact with the Palestinian ethnic identity. Khadija's mother screams and curses America when she catches Khadija with a sensual picture. At this moment, Khadija's mother comes face to face with the fact that her daughter is acknowledging her own sexuality. This moment is the pinnacle of an earlier moment in the novel when the mother insists that "no daughter of mine is American" and incidents that the mother–daughter bond breaks if the daughter defines herself as "American." In the Freudian family structure, the father figure breaks the mother–child bond so that the child has to identify with the father and the "symbolic order." Here, the father figure and Americanness play the same role in severing the mother–daughter bond between Khadija and her mother. To become a woman, Khadija has to identify with the phallic law that renders her sexuality invisible and sever ties with the maternal. By the same token, to become "American," Khadija severs ties with her mother, and metaphorically with the Palestinian tradition. When Khadija plays the maternal role, taking care of her siblings when her mother travels to the West Bank to visit her ailing mother, Khadija reports her father to the authorities for beating her brother, an act that is not expected from her. Thus, Khadija breaks the bond with the father figure as well. The novel ends without telling what happens next. This opens up several possibilities for Khadija whether remaking bonds with the father figure/mother figure or simply not bonding with either. Khadija's *feminist* stance implies her rejection for what is deemed traditional values and "non-contemporaneous."

Unlike Khadija, Hala and Mawal sustain the mother–daughter bond, in the sense that they tend to extend their mothers' interest in keeping stories and connection with their nation of origin, West Bank/Jordan. Hala always recalls her mother's advice to her: "Remember for yourself and for your tomorrow" (218). In a similar sense, Mawal and her mother are so close that people in Nawara would consider them sisters and share their stories freely with them. To recall Harootunian, Hala and Mawal live in different but coexisting temporalities, "where fragments of the past…impinge upon the present" (42).

In conclusion, non-national time in *West of the Jordan* is formed at the conjunction of the contradictory social and cultural structures between Nawara and Glendale, rooted in what is deemed traditional and precapitalist versus what is deemed contemporary and modern. This contradiction is allegorized by the complicated embroidery on rozas, "with both Palestinian and western stitches and patterns." This contradiction dismantles a homogeneous move in time for the Palestinian diaspora either in Nawara or in Glendale.

The Last Generation

The Last Generation (1993) is Moraga's second book, after *Loving in the War Years* (1983). Together with her third book, *Waiting in the Wings: Portrait of a Queer Motherhood* (1997), the three texts form a trilogy of Moraga's journey of self-identification as Mexican, Chicana, feminist, activist, and lesbian mother. Best known for her coedited feminist critical anthology titled, *This Bridge called My Back: Writings by Radical Women of Color* (1981) with Gloria Anzaldúa, Moraga writes more than one genre. She is an essayist, a poet, a playwright, and a critic. Moraga's writings are deeply personal and highly political. Throughout her narrative quest to identify her sexuality, Moraga criticizes sexism and homophobia in the Chicano nationalist movement of 1968 as well as in US mainstream patriarchy. To that end, Moraga's feminist critique draws on indigenous folktales and mythical figures in Chicano/Mexican cultural traditions to reconfigure Chicano nationalism, on the one hand, and revisit the prevalent heteronormative ideology of nationalism. Hence, Moraga proposes revisiting the notion of the indigenous nation Aztlán and suggests "Queer Aztlán" as a more inclusive nation. Along these lines, Moraga undermines the traditional heteronormative family structure that haunts the concept of the nation through her nonconventional lesbian motherhood. Intertwined with her critique of patriarchy, sexism, and homophobia, Moraga, in addition, relates the usurpation of the Mexican territories in 1848 (by virtue of the Treaty of Guadalupe Hidalgo between the United States and Mexico) to the continued neo-imperial policies of the United States and its economic and political interventions in Latin America and the Arab-Gulf region. With this context in mind, Moraga in *The Last Generation* offers a theoretical framework for understanding "Queer Aztlán" and responds to global political events of the early 1990s.

Published in 1993, ten years earlier than *West of the Jordan*, *The Last Generation* has received wider critical attention. Most of the reviews and scholarly studies have focused on Moraga's revisiting the mythological notion of Aztlán[13] and the mythological figure Coyolxauhqui[14] to highlight the cultural specificity of Chicano/a indigenous history in the United States and to gender that history by placing sexuality at its center. As such, Moraga challenges Anglo-centric nationalism and Chicano phallocentric cultural nationalism. Race and sexuality intersect in Moraga's revisited narration of the nation. My analysis of Moraga's text echoes many of these ideas but adds a new layer by interpreting Moraga's text as an interruptive narrative to the claimed homogeneous narrative of American national time.

Moraga dedicates her book to the legacies of Audre Lorde and César Chavez as well as to "the yet unborn." This dedication has multiple implications. On the one hand, it implies the acknowledgment of and carrying on the struggle of US lesbian feminists of color (Lorde) and fights for equal rights of oppressed, exploited Mexican workers (Chavez). On the other hand, it is also dedicated to future generations. In the introduction, Moraga describes her book as "a journey into the *past* as it is into the *future*, a resurrection of the *ancient* in order to construct the *modern.* It is that place where *prophecy* and *past* meet and *speak to each other*" (3 emphasis added). This hybrid timeframe is echoed in the structure as well. *The Last Generation* is a multigenre book of poems, stories, and essays. Moraga wrote the poems over a period of seven years beginning in 1985 upon returning to Aztlán[15] (2). She specifies that *The Last Generation* is a testimonial written to mark the passage of 500 years after the Spanish conquest of the New World and written in response to the political urgency of the first three years of the 1990s. That is to say, *The Last Generation* is written at the intersection of colonization and imperialism; Moraga situates the colonization of indigenous Chicano lands that began five centuries ago with the Spanish conquest within the context of the political events triggered by the imposition of US dominance over different regions in the world. As Moraga puts it, "Chicano Nation is a mestizo nation conceived in a double-rape: first, by the Spanish and then by the Gringo" (153). The double rape dismantles a linear genealogy.

Consequently, structurally, Moraga's heterogeneous narrative does not have a "plot" in the literary sense of the term. The absence of a plot implies that there is no specific order of incidents; each of the five parts has a different theme except for the last two sections ("The Breakdown of the Bicultural Mind" and "The Last Generation") that are more coherent and connected. Despite this seemingly arbitrary structure, three main themes recur throughout the book: Moraga's concerns about Chicano cultural heritage, American imperialism, and the politics of her sexuality. Throughout, Moraga criticizes US domestic and foreign policies that have contributed to wars, unfair economic policies, and negligence of the impoverished. This heterogeneous structure opens up the possibility of a nonsequential narration of the nation and proposes a new sense of genealogy for the formation of the United States.

The revisited notion of the nation that Moraga's narrative proposes intertwines the colonial history of Chicano Aztlán with the contemporary imperial history of the United States. *The Last Generation* "begins at the end and moves forward" (4); it is a testimonial about the conquest of the native lands of Chicanos—Aztlán—by the Spanish and the US seizure of Mexican land in 1848 (by virtue of the Hidalgo Treaty). On the one hand, Moraga highlights Chicanos/as' past: "we live daily in the moment of that highway robbery" (110). On the other hand, she underlines how Chicanos/as "witness [their] past in order to reconstruct a future" and they "[look] backward in order to look forward" (70, 190). The past and the present are at the core of Moraga's revisited national discourse of "Aztlán." Unlike the Chicano national discourse of "Aztlán" as a land of unity and liberation, Moraga's reconceptualization of it as "Queer Aztlán" puts gender and sexuality at its center to resist patriarchy, homophobia, and imperial exploitation.[16]

"Queer Aztlán" is a trope for a new genealogy of the nation; it is a counter narrative that interrupts the colonial/imperial story of the US nation-state. Revisiting the narrative of the US nation dismantles the paradigmatic story of Americanness founded on consensus and homogeneity, as well as the notion of an integrated American entity that is self-contained and exceptional. Moraga's narration of the US nation-state is told not in terms of its exceptionalism and isolationism but in terms of its engagement with the rest of the world and in terms of its capacity, or lack thereof, to "embrace *all* its people, including its jotería [queer]" (Moraga 147). I connect the new national temporality that emerges from queer motherhood in Moraga's narrative of "Queer Aztlán" to Julia Kristeva's concept of "monumental temporality" in "Women's Times." Kristeva argues that "monumental temporality" is "all-encompassing and infinite like imaginary space that the very word 'temporality' hardly fits" (191). This definition of monumental time resonates with Moraga's definition of Queer Aztlán as a nation encompassing all its people without imposed hierarchies. Thus, the patriarchal, heteronormative family structure is replaced by queer motherhood in the "Queer Aztlán" narrative. Anne Mclintock, in "*'No Longer in a Future Heaven': Gender, Race, and Nationalism*" (1997), holds that "nations are symbolically figured as domestic genealogies" (91). These genealogies are hierarchical; in these genealogies, women are represented as the "authentic body of national tradition (inert, backward-looking, and natural), embodying nationalism's conservative principle of continuity"; men "represent the progressive agent of national modernity (forward-thrusting, potent, and historic), embodying nationalism's progressive, or revolutionary, principle of discontinuity" (92). Queer motherhood disrupts the traditional symbolic genealogy of the nation. In her introduction to *The Last Generation*, Moraga reveals her concern about the loss of the indigenous heritage of Chicanos/as because she is unable to pass that heritage to her offspring: "My line of family stops with me." She is queer and will not have children through a heterosexual family: "There will be no one calling me, *Mami*, *Mamá*, *Abuelita*... I am the last generation put on this planet to remember and record" (9). Moraga positions herself as the last generation of Chicanos/as storytellers and heritage keepers. Nevertheless, Moraga extends her cultural heritage "*through* queerness, rather than in spite of it" as Lisa Tatonetti explains (241). Moraga revisits the traditional patriarchal family structure in two ways. First, Moraga foregrounds her perspective as a daughter and a potential mother rather than foregrounding the perspectives of the male members of her family. Second, there is a resonance between Moraga's perception of herself as an empowered daughter and the mythical goddess Coyoloxauhqui. The patriarchal story of Coyoloxauhqui revolves around the bad daughter who wants to kill her mother upon knowing of the birth of Huitzilopochtli, her brother. Moraga reverses this story and celebrates the daughter figure. Shifting the perspective through which this myth has been known symbolically restores the powers of the goddess Coyoloxauhqui.

Moraga also dismantles the patriarchal domestic genealogy in her portrayal of the father figure as possibly gay. However, the portrayal of Moraga's father is problematic. Although he is the dominating father, and although Moraga's perspective as a daughter and a mother is at the center, the perspective of Moraga's own mother is overshadowed, perpetuating the representation of the mother as a subordinate figure. Another layer of complexity unfolds, especially in the chapter titled "The Breakdown of the Bicultural Mind," if one takes into consideration that the mother figure is brown Mexican and the father figure is white Anglo. The Mexican mother and the Anglo father echo symbolically, "the Violated Mother" and "the rapist Father" (128). That is to say, the Anglo figure conjures up two acts of betrayal in the Chicano collective consciousness: the rape of the indigenous Indian woman who became the mother of the first mestizo (Mexican of Indian and Spanish descent); and homosexuality as a betrayal of the Chicano movement (the La Raza). The effeminate, nonphallic, white father figure reaffirms the racialization of homosexuality as white in Chicano nationalism. Although suppressed, the Anglo-European heritage is an integral part of the Chicano lineage. Moraga dismantles not only the patriarchal domestic genealogy but also the notion of simple transmission of a single heritage from parents to children. Moraga's family history correlates with Chicano/a national history and reveals a contradiction in her poem "Whose Savior?" (102).

Though the poem opens with "I hate white people / white blood," later in the same poem she acknowledges her mixed-race identity as well as that of all Chicanos/as: "the white people I am" (102). Chicano/a national history is a combination of indigenous heritage, European/Spanish conquest, and US imperialism.

By the same token, there is no homogeneous movement in national time. As I have mentioned earlier, centralizing Coyoloxauhqui's perspective dismantles the patriarchal genealogy in rewriting the national narrative of Aztlán. Coyoloxauhqui is not the only indigenous female figure that Moraga's narrative evokes; she also evokes the mythical mother figure in Chicano/a heritage "*before* the 'Fall,' *before* shame, *before* betrayal, *before* Eve, Malinche, and Guadalupe; *before* the occupation of Aztlán" (72 emphasis added). Moraga rewrites the stories of the indigenous figures from a precolonial perspective to shatter the colonial patriarchal narrative. This revisited story implies two contradictory movements in time: *before* and *after* the conquest. As I have quoted earlier, Moraga points out living "*daily* in the moment of that highway robbery." The movement in "*daily*" clock time implied in this quote is opposite in direction to indigenous time; whereas clock time is unidirectional, indigenous time is cyclical.

These two temporalities are entwined and create two overlapping narratives of the US nation-state and Queer Aztlán. Moraga's reconceptualization of the native Chicano/a lands renders Aztlán as an "imagined geography,"[17] in Edward Said's sense, namely, *reclaiming*, *renaming*, and *reinhabiting* a precolonized territory (225). In order to restore the Chicano/a claim of possessing New Mexico as part of their native Aztlán land, Moraga in the poem "New Mexican Confessions" uses Spanish words for landscape description: "piñón / cañón / arroyo (pine, canon, brook)" (34). As such, Moraga defies the "epistemic violence of [Anglo-American/English-only] cultural hegemony" as Mary Pat Brady says (157). Moraga, in a word, "spanglicize[s]"[18] the American landscape.

"Queer Aztlán" is reconfigured as a point of origin for Chicanos/as that both reinforces and revisits the indigenist, nationalistic notion of "Aztlán." It becomes allegorically the ground of collision and cultural violence inflicted by Euro-American conquest and usurpation of land, US imperialism of manifest destiny, patriarchal and homophobic misogyny. Historically, Aztlán is a pre-Columbus, indigenous land. The overlap of the colonial and the imperial renders "Queer Aztlán" as a trope of "cultural hybridity" and "the ground of conflict, contradiction, change, intervention" as Kumkum Sangari explains in another context (4). Therefore, the linearity of "American" national time and of Chicano national time is replaced by a more hybrid time that does not follow a linear order. It is rather a spiral order with loops, inflicted by the aggression of colonialism.

Moraga's revisioning of Aztlán resonates with her redefinition of "Americanness" that is drawn along the lines of a pre-Columbus past on the one hand, and current political crises both within and outside of the United States on the other hand. Echoing José Martí in "Our America," advocating a "union in the continental spirit" and criticizing US hegemony over central and Latin America (120), Moraga defines her Americanness in terms of belonging to the Americas rather than belonging to the geographical boundaries of the United States. She describes herself as "an American writer in the *original* sense of the word," an "Américan *con acento* [an American with an accent]" (62 emphasis added). That is to say, Moraga reopens the definition of Americanness into an older pre-Columbus history, traversing the geographical boundaries of the US nation-state. Thus, the "'*beyond*-national' comes from a '*before*-national' semantic," to quote from Sangari—in a different context, in "Ruptures, Junctures, Returns" (167 emphasis added). This, going back to the prenational, preimperial, precolonial time, challenges the core ideas of the United States as a nation, namely, its separatism and isolationism, and creates "pockets or series of transformative eruptions," as Sangari puts it (134). Instead of a limited definition of Americanness, Moraga defines an "American" as a member of "a larger world community composed of many nations of people and no longer give[s] credence to the geopolitical borders that have divided us.... Call it racial memory. Call it shared economic discrimination" (62). Whereas technological developments in media and means of communication facilitate framing an "imagined community" of Palestinians in *West of the Jordan* in which time is not homogeneous, racism, wars, and economic discrimination have shaped an "imagined community" on a hemispheric scale in *The Last Generation* where an essentialized myth of origin becomes a critique of US imperialism.

I interpret the point of intersection between "Queer Aztlán" and a hemispheric sense of Americanness in light of what Ricoeur in *Time and Narrative* calls the "historical present." As Mark Muldoon in *Tricks of Time* explains the "historical present," it is the time in which "the past comes alive… each member of a community, each storyteller borrows from the past, and works with it toward the future…. [It] is the axial point where history both ends and begins" (212–213). The "axial point where history both ends and begins" in the chapter "War Cry" is, for Moraga, the manifest destiny discourse.[19] In this chapter, Moraga exposes the intertwined relation on the one hand between global capitalism and claims of bringing democracy to the Western hemisphere, and on the other hand the deterioration of local economies and the eruption of political unrest that have forced many to migrate to the United States. That is to say, Moraga's criticism unfolds two opposite narratives about the United States: democracy and free economy versus imperialism and labor exploitation. She writes:

> When U.S. capital invades a country, its military machinery is quick to follow to protect its interests. This is Panamá, Puerto Rico, Grenada, Guatemala… [...] Every place the United States has been involved militarily has brought its offspring, its orphans, its homeless, and its causalities to this country: Vietnam, Guatemala, Cambodia, the Philippines… (55 ellipsis in original)

Throughout her text, Moraga gives specific dates to some parts of her book while some parts are not dated. The parts that are dated are about invasion, conquest, and movements of resistance to both. The parts that are un-dated are mostly the ones where she reflects on the consequences of histories of colonization and imperialism that have affected her personal life as lesbian, feminist, and Chicana. In that regard, she relates colonization and conquest of land to bodies of women, homosexuals, and all the unprivileged in the world.

Moraga redefines the notion of the nation to be one of people "bound together by spirit, land, language, history and blood" (168–169)[20] as well as the notion of the "land" to be one at the center of which lie the histories of colonization and the struggles against it:

> For immigrant and native alike, land is also the factories where we work, the water our children drink, and the housing project where we live. For women, lesbians, and gay men, land is that physical mass called our bodies. (173)

Furthermore, Moraga genders this land and feminizes it. She writes: "Like woman, Madre Tierra has been raped, exploited for her resources, rendered inert, passive, and speechless" (173). Correlating land to sexuality and to a woman's body renders the notion of land as a "third space" that is within and between Chicano male-centered nationalism and Anglo-American nationalism; both rely on linear temporality. Within this third space nationalist notion of land, the notion of land is formed in the violent act of rape—of Malinche, mother of first mestizo—and colonization; it is "more than the rocks and trees, the animals and plant life that make up the territory Aztlán" (172–173). Reconceptualizing Aztlán as a nationalistic land of origin in terms of violence and violations runs in contrast to the notion of "land" as a newly discovered and rich land.

In that regard, Moraga makes an explicit reference to Walt Whitman in her poem "New Mexican Confession," subtitled "Upon reading Whitman fifteen years later. Jemez Springs, 1988" (34). Within the poem, Moraga evokes Whitman's "Leaves of Grass":

> I fall to sleep contemplating the body of the poet
> Whitman at my age, 100 years ago
> and see his body knew the same fragility,

> the desire to dissolve the parameters of flesh
> and bone and blend with the mountain
> the blade of grass
> the boy.
> I *bleed* with the mountain
> the blade of grass
> the boy
> because my body suffers in its womb.
> The maternal blood that courses this frozen ground
> was not spilt in violence, but in mourning. (36–37)

Moraga alluding to Whitman is not surprising since he is one of the literary founders of the "American" landscape. Yet Moraga renames the American landscape in non-Whitmanesque, nonmasculine terms. "[B]leed" and "womb" foreground the feminine as part of that landscape in which Moraga's body will "dissolve" and "blend" with the "mountain/the blade of grass." Not only does Moraga replace a male-centered perspective with a female one by using female imagery, but she adds a maternal perspective when she describes her body suffering "in its womb." Moraga's lesbianism would not allow her to be a mother in a heterosexual sense; it is a "frozen ground." However, becoming a queer mother rewrites the image of the mother figure who is either the violated mother (Malinche) or the sexless, virgin mother (Guadalupe). The mother figure, as rewritten by Moraga, is an attempt to revisit and change the term of the male-centered nationalistic Chicano narratives as well as nationalistic Anglo-American narratives.

In conclusion, *The Last Generation* and *West of the Jordan* open up the possibility of a new sense of time pertaining to the formation of the nation in terms of imagined transnational and/or dispersed communities. The nation's time is reconceptualized from the perspectives of the migrants, the displaced, and the colonized. Read together, *The Last Generation* and *West of the Jordan* unfold different types of nonhomogeneous movements in time, which I argue as queer and non-national, respectively, and juxtapose times of linearity and imperialism that characterize the history of the United States.

CHAPTER 3

Moments of (Un)belonging: The Spatial Configuration of Home(land) in *The Time between Places: Stories that Weave in and out of Egypt and America* and *The Namesake*

A revisited notion of "imagined (transnational) communities," I argue, is the ground for reconceptualizing the interpellation of the *ethnic* subject in the United States, which changes the relationship between the communities imagined to form the nation and their movement in time. Thus, the consciousness formed by "imagined (transnational) communities" emerges within a "simultaneous" but "non-synchronous" temporality—in light of Bloch's theory—rather than within a homogeneous, seemingly empty time as Anderson holds. These changes are intertwined with changes in the spatial perception of the nation as a *trans*national space that blurs the familiar correlation between territorial demarcations and national affiliations. In this chapter, I argue that a new reimagining of national space that goes beyond the confinements of geographical boundaries complicates, on the one hand, what counts as home(land) perceptually and physically. On the other hand, I elaborate that within this new spatial imagining of the nation, there is a non-national moment that problematizes the notion of home(land) and diffuses political allegiance for the *ethnic* subject in both the country of origin and residence.

US migrant narratives that portray the experiences of early European settlers, especially during the early to mid-nineteenth-century decades of "literary nationalism"[1] emphasize a sense of Americanness, defined spatially as vast, new, and open. Nevertheless, as Dalia Kandiyoti points out in *Migrant Sites: America, Place and Diaspora Literatures* (2009), in "classical American literature" the spatial imagination of the United States reveals racialization and enclosure.[2] By the same token, Avtar Brah in *Cartographies of Diaspora: Contested Identities* (1996) suggests reconfiguring the United States as a "diaspora space"[3] rather than as a "nation of immigrants" (205). In this chapter I suggest an understanding of the nation as a space of suspension and ambivalence in which the preset perception of homeland and belonging is dismantled and replaced by the notion of home mediated through mobility and integration, or conversely, exclusion. To that end, I focus on the setting of the events in the stories of migration depicted in *The Time between Places: Stories that Weave in and out of Egypt and America* by Pauline Kaldas and in *The Namesake* by Jhumpa Lahiri.

Setting is important to the analysis of both books because mobility and dwellings are central to the stories and the experiences of migration, and by extension the redefinition of Americanness. Nevertheless, it is not only Americanness that is redefined in *The Time between Places* by Kaldas and in *The Namesake* by Lahiri, but the sense of belonging to a homeland is also reconfigured. The stories in *The Time between Places* are set in Egypt and the United States to portray the feelings of different characters about the decision to migrate to the United States and to show what the United States is like in their minds, as well as their feelings about the decision to return to Egypt and their perception of how Egypt has changed during the years of their migration. Along the same lines, *The Namesake* depicts the perspectives of the Gangulis, especially Ashima (the mother figure) and Gogol (the son of Ashoke and Ashima), about fitting into their surroundings (neighborhoods, apartments, houses, as well as with families, friends, and coworkers). In my analysis of the two texts, I do not focus on "home" as a physical place but on the feeling of at-homeness (or lack thereof) that the experience of migration provokes.

In this regard, I borrow from Susan Strehle's distinction, in *Transnational Women's Fiction: Unsettling Home and Homeland* (2008), between *houses* and *homes*. Strehle argues that "While houses, apartments, and other dwelling places are tangible, concrete spaces, 'home'... [is] loaded with emotional and ideological investment" (5). Feelings of at-homeness in *The Time between Places* and *The Namesake* emerge from social relations and networks, or to refer to Anderson, a shared sense of community. Building on Anderson's definition I would add that communities are formed emotionally and ideologically. Nevertheless, in light of my exposition of "imagined (transnational) communities" and the interrupted interpellation process (emotional and ideological), social relations and networks are formed beyond the limited confinement of a singular nation and the notion of the nation is reconfigured. In this sense, I borrow Rob Wilson and Wimal Dissanayake's conceptualization of the nation as an "imaginary state" in *Global/Local: Cultural Production and the Transnational Imaginary* (1996). In the era of globalization, Wilson and Dissanayake contend, the nation is "the as-yet-unfigured horizon of contemporary cultural production by which national spaces/identities of political allegiance and economic regulations are being undone and imagined communities... are being reshaped" (6). Along these lines, I suggest that the concept "imagined (transnational) communities" does not only imply the formation of a new kind of consciousness and nonhomogeneous movement in time but also, quoting from Anthony Giddens's *The Consequences of Modernity* (1990), fosters "relations between 'absent' others, locationally distant from any given situation of face-to-face interaction" (18). Communities formed among *locationally distant* members have a *constructed* "now" moment, which as Doreen Massey explains in *Space, Place, and Gender* (1994) provides "a source of dislocation... for people are everywhere conceptualizing and acting on *different spatialities*" (4 emphasis added). It is precisely at this moment of *dislocation* that I situate the reconfiguration of home(land) as a "non-national space." By dislocation I mean not anchored in either the home country or the host country, but suspended between both. My definition of "non-national space" resonates with Nadji Al-Ali and Khalid Koser's redefinition of "home" in *New Approaches to Migration: Transnational Communities and the Transformation of Home* (2002). Quoting Roger Rouse, Al-Ali and Koser hold that "home has become a moveable concept, it is pluri-local…a community created within the changing links between here and there" (6). Al-Ali and Koser observe that this new sense of home is "accompanied by a sense of rupture and discontinuity" (9). *The Time between Places* and *The Namesake* reveal different nuances of dislocation and discontinuity evoked by the experience of migration. In *The Time between Places*, feelings of displacement and rupture occur, almost simultaneously with the moment of making the decision to leave, even before actually migrating. In *The Namesake*, home is multilocated, yet there is a perpetual tension between feelings of settlement and unsettlement. Paradoxically, this feeling of (un)settlement is shared by Ashima, who has actually experienced migration, and Gogol, her son, who has not migrated but has *inherited* a sense of unsettlement through his migrant parents.

Theoretical approaches that have focused on the concept of "home" and settlement reveal the complexity of the notion of home(land). Studies by theorists such as Edward Said, Amy Kaplan, and Donna Gabaccia, to name only a few, underline the correlation between home and nation. Said, in *Reflections on Exile* (2000), emphasizes that home is a community of language, culture, and customs that affirm nationalism and nationalism is an extension of the settled home. By the same token, Kaplan, in "Manifest Domesticity" (1998), holds that home evokes the political agenda of the nation. Thus, migrants are the foreigners within and, hence, "unhomed" (111). Unlike Kaplan, Gabaccia, in *From the Other Side: Women, Gender, and Immigrant Life in the U.S., 1820–1990*, maintains that migrants blend the old and the new in "idiosyncratic ways" (114) to create an ethnic American domesticity. Other theorists such as Susan Strehle, Doreen Massey, and Homi Bhabha reconceptualize *home* as a *political* space. Strehle criticizes forms of analysis that depoliticize *homes* and inscribe them as sites for the emotional, the spiritual, and the familial rather than "places where worldly power is negotiated on a daily basis" (18). In a similar sense, Massey in *Space, Place, and Gender*, contrary to the views that see space as stasis and depoliticized, views space as inseparable from social relations and concludes that the spatial is "both open to, and a necessary element in, politics" (4). Massey's and Strethle's insights resonate with Bhabha's definition of "unhomeliness" in *The Location of Culture* (1994), as "the estranging sense of the relocation of home and the world" (13). In the experiences of relocation caused by migration, exile, and/or colonialism, "the borders between home and the world become confused," hence, "recesses of the domestic space become sites for history's most intricate invasions" (13). In light of these views, I argue that non-national space is a mobile, fluid space that opens up different kinds of political engagements formed by virtue of the experience of migration. My analysis of the lives of migrant families portrayed in *The Time between Places* and *The Namesake* reveals that these families are not anchored in specific territories or homes; hence, their political and national affiliations are replaced by local and/or regional affiliations as they struggle to be fully integrated in a national homeland.

The Time between Places

Since its publication in 2010, *The Time between Places* has not received any critical attention, neither have Kaldas's earlier two books of poetry: *Egyptian Compass* (2006) and *Letters from Cairo* (2007). However, *Dinarzad's Children: An Anthology of Contemporary Arab American Fiction* (2004), which Kaldas has coedited with Khaled Mattawa, has received some reviews.[4] These reviews mostly outline the main themes the anthology is addressing and highlight what the anthology can attain in the literary arena of ethnic/American fiction: familiarizing more readers in the United States with Arab-American literature, underscoring the heterogeneity of Arab Americans, depicting the burden of migration which

Arab Americans share with different migrant groups in the United States. This lack of attention can be attributed to the complexity of categorizing the racial identity of Egyptians as Arabs. Egypt joined the Arab League in 1945, when it was under British/Ottoman colonization. After the 1952 revolution and under the Nasser regime, a pan-Arab, pan-African, pan-Muslim nationalist discourse was emphasized more. Sadat's era, after Nasser's, witnessed a shift and a revival of the Pharonic, Mediterranean heritage of Egypt as well as an Islamist (rather than pan-Arab/African) rhetoric.[5] Although Nasser is valorized in the Arab region, especially in countries that are directly involved in the Arab-Israeli conflict (such as Palestine, Syria, Jordan, and Lebanon), many Christians fled Egypt for fear of political wars and religious oppression. As I will show in my analysis of *The Time between Places*, many of the stories are about Egyptian-Christian families who decided to leave Egypt under the rule of Nasser.

As the title suggests, the stories in *The Time between Places* weave in and out of two places: Egypt and the United States. That is to say, place is at the center of this text. The centrality of place is significant in the stories, in the dream at the opening of the book, as well as in the phone calls between a mother in Egypt and her daughter in the United States. Whether in the dream, the phone calls, or the stories, characters and places are closely tied. This intertwined relation between characters and places conjures up Massey's conceptualization of space as "social relations 'stretched out'" (2) and of place as "a particular moment in...networks of social relations and understandings" (5). By the same token, Massey elaborates that the spatial "is integral to the production of the social, and not merely its result" (4). In this sense, the perception of the United States and Egypt in *The Time between Places* is constructed out of multiple social relations between characters inside Egypt, the United States, and across both countries.

In the dream that opens the book, the mother and the daughter are holding hands and walking in "the sea in Eskenderia [Alexandria]" (2). The mother has left the daughter behind on the other side of the water, time has passed, and the distance between them has grown. As the mother moves away from the daughter, the daughter sees her mother "like a dot on the horizon" (2). The spatial description of the mother—like a dot on the horizon—implies the growing distance and space between the mother and the daughter. This image resonates with the phone calls between them as well as with the stories in the book that revolve around characters leaving, staying, and/or going back to Egypt or to the United States That is to say, space and place are thematic and structural axes in *The Time between Places.*

The growing distance between the mother and the daughter, and the daughter's fear of being left behind are feelings that are echoed in almost every story either in the portrayal of characters in Egypt imagining their future in the United States or in the portrayal of characters in the United States looking back at their lives in both Egypt and the United States. Moments of *looking forward* and *looking back* in the stories are characterized by ambivalence and/or emptiness as I will clarify with examples later in the chapter. Structurally, the four parts of *The Time between Places* are sequential. The book has a preface and the titles of the four parts chart a linear experience of migration: premigration hopes and worries and postmigration adjustments and concerns. The preface highlights the book's depiction of "different characters *inside* the experience of emigrating from Egypt to the United States" (ix). To portray emigration from an insider's perspective entails a "thick description"[6] of its intensity, which Kaldas attains by arranging each group of stories to reflect specific moments such as leaving, coming to terms with a different culture, returning to Egypt, and reflecting on the whole journey of leaving a familiar place for a new place. The characters in the stories can be metaphorically described as dots connected to form a bigger landscape of migration. While the theme of migration "stretches across the progression of these stories," as Kaldas says in the preface (ix), it is also encapsulated in the five phone calls that are distributed across the four parts of the book.

The phone calls are between the mother in Egypt and her migrant daughter in the United States, exchanging greetings during Egyptian Christian feasts and news about members of the family. In these phone calls, readers know only what the daughter says and, indirectly, know what the mother says through the answers of the daughter: "yes, we're well"; "yes, the phone lines were open" (4,70). The sentences are shortened to phrases. In addition, there are repetitions of "Aloo-Aloo-mama-Aloo" (4) and questions about father, sister, and uncles to indicate that these are long-distance phone calls. Thus, the narrative technique of the phone calls is neither showing (including the actual conversation between the mother and the daughter) nor telling (describing in details how the characters feel about the phone calls). For example, in the third phone call, the daughter receives the news of her father's passing away; the scene is not portrayed in a dramatic way, though it is directive: "What—baba—no—how—his—heart—as soon as he woke up—it can't be true—no—how—mama..." (112). This indirect communication between the daughter and her family members is significant in two ways. First, it shows that the daughter does not have direct access to any of the incidents happening to her family in Egypt. Second, because of belonging to different time zones, there is a time lag between the time an incident takes place in either Egypt or the United States and the time the daughter and the mother share the news. In other words, despite the immediacy in communication that the telephone can offer, the telephone in *The Time between Places* is dysfunctional in that regard. The phone calls are written, not as full sentences, but as incomplete phrases separated by dashes "—". These dashes function as ellipses in the conversation between the mother and the daughter that metaphorically stand for the different spatial realms they occupy. Recalling Massey's definition of the spatial as inseparable from the social, the ellipses imply not only that mother and the daughter occupy different spatial realms but also mark a significant change in their interaction, which can be exemplified by observing two different religious cultural traditions: Eastern Orthodox Christian and Western European Church.

Telephone contact among migrants, diasporas, and exiles is crucial as Hamid Naficy holds in *An Accented Cinema: Exilic and*

Diasporic Filmmaking (2001); telephoning "is instantaneous and simultaneous. Its live ontology *obliterates* spatial and temporal discontinuity" (133 emphasis added). Contrary to what Naficy argues, telephoning in *The Time between Places* emphasizes, rather than obliterates, the time difference between Egypt and the United States. Unlike telephoning in *West of the Jordan*, around which an imagined community is formed among Palestinian migrants and their family members in Jordan and the West Bank, the phone calls in *The Time between Places* are mostly about general subjects (such as the time difference between Egypt and the United States, what the daughter and her family eat for Christmas/Easter, observing Christmas/Easter according to the Coptic calendar or the Western European Christian calendar). Despite the convenience of making phone calls between Egypt and the United States, the continuous reference to the time difference between both places and how the daughter manages to avoid jammed phone lines, illustrate the difficulties of maintaining transnational affiliations. That is to say, the phone calls in *The Time between Places*, marked by ellipses, form an emotional narrative of dislocation that reveals the inability of the migrant daughter to belong fully to either Egypt or the United States. I analyze these elliptical spaces in the migrant narrative which *The Time between Places* portrays as moments in which the notion of *home(land)* is reinscribed as a non-national space.

Non-national moments in *The Time between Places* are moments and feelings of uncertainty that characterize the spatial imagination of the United States and/or Egypt from the perspectives of different characters. These moments recur in every group of stories whether the stories take place in Egypt or in the United States, before or after migration. Though the four parts of *The Time between Places* are organized thematically, I suggest reading the stories structurally, arranged around three moments: imagining the United States as a prospective homeland; imagining, in retrospect, life in both the United States and Egypt; and imagining Egypt as a prospective homeland for Egyptian returnees. These three moments suggest that in the migrants' minds, two nations are simultaneously configured and blur demarcations of national belonging based on geographical boundaries. I, therefore, examine two trajectories for non-national moments in *The Time between Places*: spatial imagination home/nation, and the characters' sense of belonging.

Mustafa in "A Game of Chance" (that falls in the first part, titled, "Chance Departures") works in a pastry store, wrapping pastry packages; he looks for better chances in the United States. Noticing the customers of the pastry shop, Mustafa categorizes them according to their gender, citizenship, and consequently the tip each would leave him: men are "consistent" with their tips but women are "erratic"; foreigners are inconsistent with their tips (the ones who know some Arabic are less generous than the ones who are tourists or who are temporarily in Egypt). However, Mustafa finds it difficult to categorize the Egyptian woman who has lived most of her life in the United States. She is reasonably generous and she is "not skittish like the foreigners and not frantic or tired like the Egyptians" (7).

The Egyptian-American mother who comes to the pastry store with her two daughters intrigues Mustafa about what it is like to live in the United States and whether his son will learn and speak in English to him if they live in the United States. That is to say, though Mustafa thinks of the United States as the land of opportunities and better living conditions, it also evokes a sense of ambiguity in his mind. This sense of ambiguity turns into uncertainty and anxiety when he pictures himself in the airport and ready for taking off: "He saw himself *suspended* in the sky, *unable to imagine his landing*" (20 emphasis added). This paradoxical moment of imagining the inability to imagine landing preempts his perception of belonging to either Egypt or the United States. Thus, hopes and dreams of new territories and opportunities give way to fears of rootlessness. Feeling suspended is a moment in which "the allure of escape and the pull of the permanent rub against each other," to borrow Naficy's words (243). I interpret this standstill moment as a non-national moment that emerges from the anxiety of encountering what Wilson and Dissanayake call "the *as-yet-unfigured* horizon" (6). Though Wilson and Dissanayake use the term to describe the synergy of the local and the global in contemporary global capitalism that dismantles national affiliations, I borrow their term here to point out the intricacies of shifting affiliations and forming new *imagined transnational communities* from which the perception of the nation as a non-national space emerges.

Mustafa's perception of himself is ignited by his perception of the United States as the land of opportunities that is open to any hardworking person who wants to pursue his/her dreams. Nevertheless, when the staff member who conducts the interview with visa applicants in the American embassy says, "America has the right to choose who to let into its country" (7), this casts doubt on what the United States seems to promise and what it actually delivers.

The first group of stories is about the intriguing possibilities the United States can offer and the anxieties that accompany the migrants' hopes, as they tread an unknown terrain and take chances. The second part of *The Time between Places*, titled "Early Arrivals" depicts Egyptian migrants in the United States. Every story in this part portrays incidents occurring in a specific place that symbolically stands for the characters' experiences of migrating to the United States and their perception. The two stories I focus on are "Airport" and "The Top." In these stories the nation is perceived spatially through the daily lives of the characters.

"Airport" is a story about Samir in the United States and Hoda in Egypt, whose marriage was arranged by Samir's brother and Hoda's parents. Samir did not see Hoda; his brother proposed on his behalf, and the marriage is to take place upon Hoda's arrival in the United States. The story is narrated alternatively from the perspective of Samir while waiting for Hoda in the airport and from the perspective of Hoda while packing her suitcases to take to the United States. Samir's thoughts and feelings of anticipation resonate with Hoda's, as she reflects on the way she is getting married and her trip to a different country. The spatial arrangement of the text that allows the reader to know Samir's and Hoda's perspectives separately echoes the narrative structure of the phone calls between the daughter and the mother, which forms an emotional narrative *with ellipsis* because members of the same family occupy different spatial realms and cannot fully belong to either the United States or Egypt.

The portrayal of Hoda and Samir's arranged marriage is devoid of sentimental implications: Hoda feels the pressure to marry before she turns 30 and Samir feels the same pressure because he has already turned 30. The airport is Hoda's and Samir's first-date place. Airports, in Naficy's words, "are not just rhizomatic points of linkage to other points in an abstract network of relation and commerce... [They] are *nodal sites* of high intensity in which [feelings of] belonging and unbelonging are juxtaposed" (246 emphasis added). Elaborating on Naficy, I argue that airports can be "nodal sites" in the sense that they are emotion-centered spaces, specifically for the reunion of family members. However, these implications are replaced in this story by a portrayal of the airport as an anonymous space rather than an emotionally laden space. Thus, the airport in this sense becomes what Marc Augé calls a "non-place." In *Non-Places: Introduction to an Anthropology of Supermodernity* (1995), Augé argues that "a space which cannot be defined as relational, or historical, or concerned with identity will be a non-place" (77). He elaborates, a nonplace "does not contain any organic society" (112) because it is a transit place. The airport, hence, stands metaphorically for Hoda's paradoxical reasoning for coming to the United States. On the one hand, Hoda is encouraged to accept Samir's proposal and travel to the United States because the United States "is more suited to [her] independent nature" (104). On the other hand, she accepts an arranged marriage, meeting her potential husband in an airport, which—in light of its portrayal in the story—can be interpreted as a *nonplace*.

At the end of the story, Hoda decides to take only half of her belongings to the United States and leave the other half in Egypt. This decision implies partial belonging to both the United States and Egypt as well as perpetual mobility between both places rather than settlement in either, suspended, in a way, like Mustafa, between two places. The anonymity of the airport emphasizes these implications in the sense that Hoda does not feel anchored in one specific, named place. This feeling of dislocation, as I have pointed out earlier is an aspect of reconfiguring the United States, and implicitly here Egypt as well, as a non-national space.

Another story that reveals a contradictory perception of the United States is "The Top." In this story, although the United States is imagined to be like "heaven," the material reality of the main character—Shoukry—shows otherwise. Shoukry, reluctantly, agrees to quit his job in Egypt as a civil officer and moves with his wife to the United States. Shoukry's wife perceives Egypt and the United States as opposites. While "America is heaven," Egypt "is closed" and—especially under Nasser's regime—there is "no future" in Egypt (73). Most of the incidents in this story take place in an elevator in one of Boston's skyscrapers. The elevator as a confined space with strict boundaries stands in contrast to the United States as "heaven." The implications of the United States as "heaven" (read: promised land, freedom, equality, and comfort) are reversed in the portrayal of Shoukry's job as the operator of the elevator, which contrasts with his job in Egypt. The detailed descriptions of Shoukry's office in Egypt and the elevator he operates in Boston contradicts Shoukry's wife imagining the United States as the land of equal opportunities. In Egypt, the desk in Shoukry's office is

> very large. Even if he [bends] forward and [stretches] his arms to either side as far as he [can]...his fingertips [will] barely grip the edges. The desk chair [is] cushioned, and it [twirls] around. (71)

The description implies that Shoukry has an education that brings him a good job and social status. In the United States, Shoukry's job is "like a continuous circle but flat, never curving out, no interior space inside the lines," and Shoukry is described as "confined in this elevator, pushing buttons, taking people up, down. The high metal stool to lean against, occasionally rest the weight of the body, back stiff, supported by a thin slip of air" (74). Confining, hollow, and tedious describe Shoukry's world in the United States versus the large office and cushioned chair in Egypt. By explicitly challenging the view of the United States as the land of opportunities, the story writes back to the American dream as well as casts doubt on the fact of equality in the United States. With no college degree or prior experience, Shoukry's wife gets a better job than his. Not only that, but Shoukry's wife also gets her job through her sister rather than an application process.

Like the phone calls, the elevator conversations form an *elliptical* narrative. Because elevator rides take little time, Shoukry is not and cannot be engaged in a conversation with everyone in the elevator. Elevator conversations are brief, general, and mainly occur to pass the few moments spent within that space. Thus, most of Shoukry's daily conversations at work are limited to the desired floor and daily greetings. This is very similar to the phone calls between the daughter and her mother in which they exchange greetings and ask briefly about each other's lives. Unlike the phone calls that are between two closely related family members, Shoukry's elevator conversations are with strangers who do not share much with the elevator operator. The hollow space that encompasses the elevator is echoed in the empty, incomplete conversations Shoukry has while doing his job.

The elevator also recalls the anonymous airport in "Airport." Unlike the spaciousness of the airport, the elevator is a more contained space, even a claustrophobic space; yet, it is another example of the nonplace. An elevator, to clarify, is a nonplace in the same sense that Fredric Jameson explains "postmodern hyperspaces" in *Jameson on Jameson: Conversations on Cultural Marxism* (2007). The elevator is a form that "has no internal meaning of its own" (126). Jameson, in *Cultural Turn: Selected Writings on the Postmodern 1983–1998* (1998), holds that the elevator is an example of *postmodern hyperspaces* that evoke a spatial experience that challenges the subject's capacities "to locate itself or to organize its immediate surroundings perceptually, and to map cognitively its position in a mappable external world" (15–16). Though Jameson theorizes "cognitive mapping" within the wider context of postmodernity and late capitalism, I borrow this concept to interpret Shoukry's dilemma of *locating* himself or *mapping cognitively* his position after he migrates to the United States. For example, the times when

Shoukry craves a specific Egyptian meal/dish, evoked by a smell he comes across in the elevator, imply an imaginary belonging to a space outside the elevator, Boston, and the United States. In the last scene of the story when he craves *koshari*,[7] gets out of the elevator, and follows the smell that incites the craving, Shoukry, in his mind, blurs his life in the United States with his previous job in Egypt. While imagining that he is in Egypt, issuing building permits, Shoukry finds himself surrounded by an angry crowd, screaming at him and calling him "bastard," "foreign idiot," and "crazy" (79). In that moment, Shoukry is perceived as a stranger who is "not even speaking English" (79), which means that even a brief question about what is happening to him is not possible since he is, presumably, unable to communicate in English. I therefore conclude that this impasse moment of Shoukry's failing to *cognitively* locate his position dismantles the stereotypical spatial perception of the United States as "heaven" or land of freedom, and is an aspect of a different perception of the United States as non-national space.

In the first part of *The Time between Places*, Mustafa feels suspended between two places, unable to land in either Egypt or the United States. In the second part, in a similar sense, Hoda and Shoukry feel dislocated, unable to clearly perceive where they belong. In the third part, the main characters in "A Conversation and He had Dreamed of Returning" experience feelings of problematic belonging(s), however, not as migrants but rather as returnees. These two stories unfold a disjuncture between the expectations of returnees and the reality of returning. In this part, it is not the United States (as a prospective home) that is imagined but Egypt (as an originary homeland).

In the story "A Conversation," an anonymous couple—husband and wife—are arguing about whether or not they should go back to Egypt after living in the United States for 40 years. Although within the course of the story we learn that this is a Christian couple, their anonymity frustrates the assumption that names are indicators of religious identity in Egypt, which became more common under Nasser's regime and afterward under Sadat's rule. Readers would not know about the religion of the characters until the wife indicates their religious identity. The imposed Arab/Islamist identity that has dominated Nasser's and Sadat's rhetoric has suppressed the diversity of the Egyptian people and provokes a rethinking of an imposed national categorization of Egyptians.

The indefinite article in the title of the story suggests that this is a common conversation that can take place among members of migrant families in the United States. As with the anonymous characters, the story goes beyond the specific experience of the couple to a shared experience among migrant families, namely, the tension between looking back and the wish to return to one's homeland, and the possibility—or rather the impossibility—of fulfilling this wish. The conflicted perceptions of the "old" place and the "new" place in this story are set in the form of a binary opposition. The story is structured in the form of a conversation: one point of view in one paragraph followed by another paragraph with the opposite point of view. That is to say, the paragraphs alternate between the husband who wants to return to Egypt and the wife who insists on staying in the United States. Ironically, it is this same binary structure that dismantles arbitrary oppositions between home country (as a place where one ultimately belongs) and host country (as a temporary place to reside in).

Not only are the characters anonymous, but also it is not clear where they are in the United States. The only place that the husband describes is their house, and he describes it as an "empty space, the walls turning their corners, tucking [them] inside their angles, keeping [them] cloistered" (116). This description of the house as an "empty space" reflects his perception of the US nation-state as an inhospitable "empty space." To recall Massey's definition of the spatial in terms of social relations, the description of the house as an "empty space" resonates with the husband's criticism of his social life in the United States. He says:

> I have learned their language, their slang, their clothing, how to eat their food, how to laugh at their jokes, how to make their money. Still they *grimace* when they meet me, they scratch their heads instead of shaking my hand. (117 emphasis added)

The sense of emptiness that the husband implies here is an *affect* evoked by gestures of disapproval and being unwelcomed despite living in the United States for 40 years. This perception stands in contrast to what the husband imagines will be their prospective life in Egypt: a place where he feels surrounded by people with whom he is familiar and knows well. If they go back, the husband wants to "pull [his] family together again and loosen the tight grip of isolation. [They] will all return to settle [their] feet into the sand and water of [their] homeland" (118). The places that the husband mentions in the story are the Red Sea, Hurgada, and Oberoi Hotel in Giza as places to live in when they return. The implication is that the husband and the wife have enough money to afford living away from the hassle and bustle of major cities and hiring individuals to work for them. Nevertheless, the husband's speculations unfold a problematic sense of return migration. On the one hand, the husband and the wife are familiar with these specific sites by virtue of their social, family, and cultural ties to Egypt. On the other hand, the fact that staying in touristic sites is always temporary and social relations are temporary as well conjures up what Augé says about "non-places," being devoid of "any organic society." To persuade his wife, the husband assures her that they "would be free to come and go as [they] please" (116). Thus, Egypt implicitly is not their final destination; neither is the United States. Hence, both Egypt and the United States become transit points, sites of coming and going. That is to say, what starts as an opposition between living in Egypt and living in the United States reveals a life of isolation in both places.

The narrative structure of the story reveals a split in the characters' minds between Egypt as a prospective homeland they return to and the United States as their current home. Whereas the husband perceives the United States as a ruthless place, the wife perceives Egypt as an unfair place to the Christian minority. Whereas the husband projects a financially comfortable future in Egypt, the wife believes that the days of Egypt's glory are gone. Whereas the husband sees Egypt as a place of settlement, family ties, and emotional bonding, the wife perceives the United States as the place where they should belong, especially her children who are born in the United States and whose belonging to Egypt is "too invisible to follow back" (119). At the end of the story, the conversation is not concluded and the tension between the poles of the binary (Egypt–United States) is not resolved. This ending implies that the conversation is still open and will recur. The anonymity of the husband and the wife also suggests that this tension could also be an internal conflict within a single mind. That is to say, the characters' sense of belonging to either Egypt or the United States is constructed by a split where they perceive themselves as a minority in both countries. The narrative structure of the story does not only reveal a split in the spatial imaginary of Egypt and the United States but also reveals a *dialogic* spatial imaginary. That is to say, in the *dialogue* between the husband and the wife, each place (Egypt/United States) is imagined and perceived in light of the other, and no final decision is made. To conclude, in this story, the non-national spatial imaginary is precisely this lack of finalization that subverts singular definitions of belonging.

No singular definition of belonging and problematic return to Egypt are present in the story *He Had Dreamed of Returning*, in which the main character, Hani, actually returns to Egypt. Caroline Brettell and James Hollifield argue that emigration does not "necessarily mean definitive departure" and return "is not definitive return" (17). Along the same lines, this story disrupts the linear narrative of migration, when Hani, the protagonist, decides to return to Egypt. Hani left Egypt at the age of 15 with his parents after his brother was drafted and died in the 1967 war with Israel. After 20 years in the United States (during which time he has married an American wife), Hani decides to go back to Egypt for good. As the opening lines of the story suggests, the terms of belonging are defined by place of birth:

> Hani had imagined his landing back on *the land that had given him birth.* He saw himself walking the streets with his head held high and his broad shoulders embracing the world *that rightfully belonged to him.* (141 emphasis added)

Hani's sense of belonging to his home country is perceived in terms of a motherland. By virtue of this sense of filiation, Hani feels that it is his right to belong to Egypt as well as that Egypt belongs to him. Therefore, when Hani decides to return to Egypt with his American wife, Nancy, he insists on using his Egyptian passport "to officially reclaim his identity" (147–148). However, upon his arrival, despite his Egyptian passport, the Customs officer asks Hani: "Why are you back?" "*Is this your country* or the great America? Haven't you forgotten *us*?" (148 emphasis added). This conversation demarcates a moment of disillusionment for Hani, in the sense that it dismantles his assumption that when he returns to Egypt, he will reintegrate unproblematically. The officer, by the same token, draws a distinct line between Hani (who has lived outside Egypt for 20 years) and "us" (Egyptians who have not migrated). This moment, as well, exemplifies what Caitríona Ní Laoire, in "Return Migrants and Boundaries of Belonging" (2011), describes as "a disjuncture between 'home' as dream and 'home' as actually experienced" (21). I argue that the non-national moment in this story lies at this point of disjuncture, where the spatial imaginary of Egypt and the United States is reconfigured.

Hani compares Egypt to the United States in terms of roots and family affiliations. In Egypt, "you were *grounded on the foundation of your family*"; in the United States, people "built themselves out of thin air *with nothing to attach them to their origins*" (145 emphasis added). Though this distinction perpetuates the stereotypical Orientalist binary between the "Western" person as free individual versus the Arab who is tied to his/her kin,[8] Hani's return with his wife to Egypt to "reclaim his identity" shatters that binary as well as the stereotype.

When Hani returns to Egypt and starts reconnecting with his family members, his relatives and acquaintances look different from the images he has of them. In addition, Hani feels unable to catch up with their conversations when they turn to "people he [doesn't] know, events he [has] not been there for, or politics he [doesn't] fully grasp" (153). That is, although physically present in Egypt,

Hani feels alienated from the lives of his family members. Not only does Hani feel alienated among his family but also among his peers in the company he works for. At work, Hani has "sensed an *odd distance* from his new colleagues" (152 emphasis added). This spatial description of Hani's feelings of estrangement from his colleagues is heightened when his director explains how the work is handled in the company: demarcating differences between Egypt and the United States—"Egypt is not America"—and referring to the United States as "your [Hani's] America" (154). This explicit assumption that Hani belongs to the United States and the United States belongs to Hani clashes with the opening lines of the story that highlight Hani's sense of filiation to Egypt. Nevertheless, it reemphasizes a social distance between migrants and stayers. Thus, Hani occupies a liminal position between being a migrant coming back and being a newcomer. What problematizes Hani's liminal position even more is that, unlike Hani, Nancy, his American wife, with great success, comes to terms with the language, cooking, daily duties and routine, Hani's relatives, and her colleagues in school. The enduring distinction drawn in "Egypt is not America" is challenged when Nancy insists on not leaving Egypt, creating a form of spatiality defined in terms of integration rather than a shared place of birth.

At the end of the story, Hani leaves his wife in Egypt, does not seek family ties, and switches to his American passport in the airport. Brettell and Hollifield assert that "dual nationality" means maintaining "a presence abroad as well as attachment to home." Elaborating on Brettell and Hollifield, I add that switching passports (and implicitly national affiliation) in this story signifies not only an interstitial moment, but also implies a moment of "rediasporization," as Anders Stefansson holds in "Homecomings to the Future: From Diasporic Mythographies to Social Projects of Return" (2004). Stefansson underlines that homecomings have unsettling consequences that returnees confront because of transformations of their homelands, which may lead to their rediasporization in their native countries (4). The diasporization of

Hani in both the United States and Egypt and the decision of his wife, Nancy, to stay in Egypt reframe the migration story as a cyclical narrative rather than a linear one.

The final part of *The Time between Places* is titled "The Silence of Memory" and it comes before the fifth and final phone call between the daughter and her mother. As I have clarified earlier, the five phone calls can be considered a pithy migrant story with a linear plot that starts with resisting the mainstream traditions of the host country and ends with a sense of compromising between mainstream cultural tradition and ethno-cultural heritage. I have also mentioned earlier that the four parts of the book too follow a linear plot. Thus, one can consider the fourth part to be the final segment of *The Time between Places.* However, unlike the phone calls and unlike the preceding three parts where the characters imagine United States/Egypt retrospectively or prospectively, this part depicts characters that are neither *looking back* to an old place nor *forward* to a new place. In all stories of this part, memories of Egypt have receded; an aura of silence, repression, and/or sorrow dominates the general atmosphere of these stories. Characters in these stories are second-generation immigrants who, as Kaldas says in an interview, try "to create a solid place" for themselves in American culture. In the preface for the book, Kaldas describes this last set of stories as about "the lives of the next generation as they negotiate the paths between their two worlds" (ix). Rather than negotiating the paths between two worlds, I argue that in this part, what is portrayed are individual stories, each with a different dilemma, not necessarily the dilemma of the clash of cultures. What these stories share is that they all end with a feeling or a moment of uncertainty and emptiness.

For example, the story *Bluebird* is about Sonya—second-generation Egyptian-American—and her husband, Rick, who have decided to postpone having children because they are burdened with mortgage payments. Though they have managed to pay off their home loan, they cannot manage to have a child. The continuous failed attempts to have a child mark Sonya's feeling of loneliness, which is echoed in the description of the house she and Rick live in:

> The house [grows] larger.... There [are] too many rooms that [sit] still, unused, except for the kitchen and the bed room....... It [is] a large colonial house...white with slender columns in the front, giving it a kind of miniature mansion appearance. But the pillars were too thin for its large structure, making it seem like the house would tip over if they were removed. The inside [is] room after room painted in clean white strokes with straight angled walls. Sharp lines, no nooks or crannies; nothing [distracts] the eye. (198)

Sonya and Rick's house is described in terms of its physical structure: precise angles, lines, colors. Furthermore, this portrayal focuses on the size of the house, which conveys the diminishing intimacy between Sonya and Rick. In this sense, the portrayal of the house resonates with Strehle's distinction between houses and homes, which I have pointed out earlier. While *houses* are "tangible," *homes* are "loaded with emotional...investment." In other words, this portrayal reveals Sonya's lack of feeling at home. By the same token, calling on Massey's definition of place as "a particular moment in networks of social relations and understandings," I argue that Sonya and Rick's house reflects a troubled marital relationship and that Sonya and Rick occupy different social spaces. The metaphor of the bluebird versus the basement in the story accentuates the incongruent spaces that Sonya and Rick occupy. While Sonya chases a bluebird to feed it, Rick spends most of his time in the basement, renovating. Whereas the bluebird never finds a nest to land so that Sonya can feed it, the basement always looks the same, despite Rick's renovations. One can draw an analogy between Sonya's yearning for the warmth of a home and the missing nest. In a similar sense, one can also draw a correlation between the confinement of the basement and the fact that Rick spends most of his time in that limited space.

In conclusion, *The Time between Places* disrupts the homeleaving-homecoming migrant narrative as I have previously pointed out. This creates new spatialities formed by virtue of shifting belongings between home and host countries. Within these new spatialities, problematic feelings of at-homeness, belonging, and not belonging emerge.

The Namesake

Jhumpa Lahiri is the first Indian American writer to win the Pulitzer prize for fiction in 2000 for her collection of short stories, *Interpreter of Maladies* (1999). Her second most popular work of fiction is her first novel, *The Namesake* (2003), which has been adapted for a Hollywood movie production, directed by Mira Nair in 2006. After *The Namesake*, Lahiri wrote her second collection of short stories, *Unaccustomed Earth* (2008), and her latest novel, *The Lowland* (2013). The central theme in these works is migration to the United States; these migrant narratives portray Indian characters leaving for the United States and the transformations they undergo to adjust and ultimately fit in the host-country. Lahiri depicts stories of migration from the perspectives of both men and women and of first and second generations to offer insights into the Bengali and/or Indian cultural traditions and customs. Lahiri, as well as Chitra Divakaruni and Bharati Mukherjee have contributed to the popularity of South Asian American fiction in the United States, yet this popularity is problematic. Although literary works by these writers allow readers, who are not familiar with Indian cultural traditions, to have access to some of these traditions, they reemphasize stereotypes of exotic traditions and gender roles. These stereotypes, as Tamara Ayesha Bhalla puts it in *Between History and Identity: Reading the Authenticity in South Asian Diaspora* (2008), "reif[y] the Orientalist binary of repressive Eastern tradition and progressive Western modernity" (29). Along these lines, Shankar and Cheung hold that "Lahiri's characters seem to imply that they live in a more de-racialized and de-classed U.S. political landscape.... This deracination of her characters and depoliticizing of the historical events... does allow for an easier consumability of Lahiri's fiction by most readers irrespective of their racial or cultural background" (xvii). My analysis of *The Namesake* in the rest of this chapter resonates with some of these critiques. However, I argue that although racial tension is not a central theme in *The Namesake*, the novel problematizes singular belongings and hence homogeneous perceptions of either the United States or India as a home(land).

The Namesake is a migration narrative, portraying two generations of Indian migrants—Ashima and Ashoke Ganguli and their American-born children, Gogol and Sonali. Though, as the title suggests, the novel revolves around Gogol who is dissatisfied with his name that is neither Indian nor American (but Russian), the novel is also about Ashima who has joined her husband in the United States after they were married. In the 32 years that Ashima has spent in the United States, she adjusts to living in the United States and approximates Indian cultural tradition to feel at home in the United States. Like *The Time between Places*, setting in *The Namesake* reveals the characters' problematic feelings of at-homeness.

Unlike Kaldas's *The Time between Places*, Lahiri's *The Namesake* is widely acclaimed (and was made into a motion picture in 2006, as I mentioned earlier). Many reviews, such as those by Himadri Lahiri, Lavina Dhingra Shankar, and Floyd Cheung, address one or more of the following themes: generational conflict between the migrant parents and their American-born children, the tension between Bengali and mainstream American cultural traditions, the analogies between Lahiri's own predicament and the dilemma of second-generation Indian Americans who identify as neither fully Indian nor fully American. By the same token, some reviews discuss why *The Namesake* is an appealing text for a wide range of readers. As Lavina Dhingra Shankar puts it, Lahiri's fiction is attractive because it is "intriguingly spicy, yet not too hot" (32). Unlike many of the characters in *The Time between Places* whose dreams of a better life in the United States are disappointed, the Indian families portrayed in *The Namesake* are middle-class, professional Indian-American families.

The novel's first chapter titled "1968" opens with Ashima feeling the weight of her pregnancy and preparing a snack for lunch.

Ashima's lunch is a "*concoction*...a *humble approximation* of the snack sold for pennies on *Calcutta sidewalks* and on railway platforms *throughout India*" (1 emphasis added). Heaviness, vulnerability, loss, and approximation describe Ashima in the kitchen in the apartment in Central Square, Cambridge, and simultaneously conjure up what buying this snack is like throughout India. The daily routine frustrates Ashima; the approximated Indian lunch is never like the original because "as usual, there's something missing" (1). Ashima's daily efforts complicate Homi Bhabha's argument in "DissemiNation: Time, Narrative, and the Margins of the Modern Nation" (1990) that "[t]he scraps, patches and rags of daily life must be repeatedly turned into the signs of a national culture" (297). For one, Ashima's lunches are to counteract her feelings of homesickness, but there is always "something missing." In that sense, Ashima echoes the migrant daughter in *The Time between Places* and the phone calls between the daughter and her mother where there is always a time lag between the mother and the daughter and a lack of immediate access to what transpires in Egypt/United States. This suggests that the daily life of Ashima becomes a sign of a "missed" national culture that she tries to replicate but with no satisfying outcomes.

The first scene is followed by Ashima giving birth to her son in the hospital. Ashima's sense of loneliness during delivery is another moment of a "missed" national culture. Like the kitchen scene, in the hospital, Ashima pictures this moment if she were in Calcutta, where she would be surrounded and comforted by many women of her family. In both scenes there is a subtext, an alternative imagined scene for what it would be like had she been in India. Both the story and the substory do not coincide and are set in opposition. This fissure between Ashima's feelings of pain and loneliness in the hospital and what it could possibly be if she were in India is a non-national moment. It is not a liminal, interstitial moment but rather a moment of grief and loss that highlights the perception of the United States as a "foreign land" (6). Thus, pregnancy and motherhood become a trope for demarcating (non)-national spaces. Ashima "is terrified to raise a child in a country where she is related to no one" (6), "alone" (33), "unmonitored and unobserved by those she loved" (6). Ashima's fear to raise her son alone unfolds her conflation of home and homeland to suggest her resistance to raise her child in a "foreign land" and consequently reveals the fissure between the national and the foreign, formed because of the absence of her family members in the United States. Alternatively, being an immigrant is described in terms of a "*lifelong pregnancy*—a *perpetual wait*, a *constant burden*, a *continuous feeling out of sorts*. It is an *ongoing responsibility*, a *parenthesis* in what had once been ordinary, only to discover that that previous life has vanished, replaced by something more complicated and demanding" (49 emphasis added). This description evokes a paradoxical sense of continuity (lifelong, perpetual, constant, continuous, ongoing) and interruption (parenthesis, waiting), which echoes the contradiction between the national (the continued "ordinary," "previous" life) and the foreign (that interrupts that continuity and renders it "complicated and demanding"). The heaviness of pregnancy and the burden of being a foreigner are individual experiences that are characterized by a feeling of solitude. Such moments of spatial or emotional solitude are what I argue to be non-national moments in *The Namesake*.

Throughout the novel, there are specific details about where the Gangulis live. The physical descriptions of their first rented apartment in Central Square as well as their house in the suburbs of Cambridge highlight their sense of belonging (or lack thereof) to where they live as well as to their surroundings. The Gangulis' first apartment is in a house "covered with *salmon-colored* shingles, surrounded by a *waist-high chain link* fence. The *gray of the roof*, the gray of cigarette ashes, *matches the pavement of the sidewalk* and *the street*. A *row of cars* parked at meters *perpetually lines one side* of the curb" (29 emphasis added). All the houses in the neighborhood are shingled, same shape and size, and "in the same state of mild decrepitude, painted mint, or lilac, or powder blue" (30). The shingles, together with the gray roof, are reminiscent of the early settlers from Europe in the seventeenth and eighteenth centuries, which reemphasizes the notion of the United States as a "nation of immigrants" and recalls the fantasy of America/the new world as an open and expansive geography that fascinated early migrants to explore and settle in this landscape. This description also evokes a sense of symmetry, order, and coherence: the gray roof matches the grayness of the sidewalk, the buildings comply with a certain range of colors (salmon and gray), and they have a similar architecture. This uniformity suggests lack of variety. In addition, the "waist-high chain-link fence" implies a sense of both containment and confinement. That is to say, Central Square, the neighborhood that has witnessed different waves of migration and antiwar protests has, paradoxically, become the site of homogeneity and confinement.[9] By the same token, there is also a contradiction between Ashima's perception of their apartment and the houses that she has watched in movies such as *Gone with the Wind* or *The Seven-Year Itch* in movie theaters in India. The apartment in Central Square is small, gloomy, dreary, too old, and not well maintained. In a similar sense, the description of the Gangulis' first house in the suburbs of Cambridge evokes a sense of order and confinement. Ashoke and Ashima's neighbors are all "Americans": the Johnsons, the Metrons, the Aspris, the Hills; and their ways of living are almost the same: "plastic wading pools and baseball bats are left out on the lawns.... Shoes are worn inside, trays of cat litter are placed in the kitchens, dogs bark and jump when Ashima and Ashoke ring the bell" (51). The implications here are: Ashima and Ashoke are the only Gangulis (Bengali family) in the neighborhood; and, consequently, they feel they are foreigners rather than neighbors. In addition, their house is described as "erected on a *quarter acre* of land. This is the *small patch* of America to which they [Ashoke and Ashima] lay claim" (51 emphasis added). Contrary to the classical image of expansion and vastness, "quarter acre" and "small patch" imply enclosure and boundaries.

Despite Ashima's feelings of disappointment and estrangement, when she writes to her parents, she still writes about the advantages of living in the United States: powerful cooking gas, four-burner stove, abundant hot and cold tap water (30). Ashima's repetition of the familiar perceptual images of the United States as a place with better living conditions heightens her feelings of loneliness and creates two paradoxical senses of ambivalence. On the one hand, Ashima is reluctant to raise her son in the United States and pities him "entering the world so alone, so deprived" (25). On the other hand, Ashima is reluctant to share her feelings of disappointment with her parents. This double ambivalence suggests a third realm that Ashima occupies that is neither part of her social space in Central Square/Pemberton Road, nor a part of her parents' cultural space.

This third realm, the site of Ashima's double ambivalence, is also the site for what I call "compensatory homes."[10] Ashima relies on a network of Bengali families to replicate cultural heritage: living within walking distance of other Indian families, meeting every weekend, participating in and attending different ceremonies. A case in point is the "rice ceremony" that marks Gogol's first bite of solid food. Approximating all the rituals, Ashima and Ashoke's Bengali acquaintances play the same roles that Gogol's grandparents and uncles would traditionally play had they been in Calcutta. Similarly, Ashima adopts American cultural traditions of Thanksgiving, Christmas, and Easter: roasting turkey with Indian spices, nailing a wreath on Christmas, coloring eggs on Easter (64). In that sense, the idea of "compensatory homes" reverberates with classic migrant narratives and the linear plot of resistance, adjustments, and ultimately mixing the old with the new. However, Ashima's frustration in her attempts to approximate and adapt Bengali and American cultural traditions make them signs, not of reconciliation, but of a "missed" national culture. This is emphasized even more when Gogol and Sonali resist adopting their Bengali heritage. Gogol and Sonali are not interested in learning the Bengali alphabet; they are more excited about American cultural occasions than the gatherings to worship Bengali deities or birthday parties for other Bengali children. Symbolically in the novel, in the "rice ceremony," Gogol does not pick any of the three options that foretell his future and his career (a landowner, a scholar, or a businessman). That is to say, Gogol occupies a different social space.

Gogol's undetermined social space is intertwined with the confusion about giving him a name. The fact that the baby boy is not given a name right after his birth, waiting for the grandmother's letter with suggested names—a letter that never arrives—suggests a *time lag* in entering the Bengali world of his parents and will ultimately lead to his shaky entry into the white mainstream cultural space. That confusion takes place when the baby boy's "pet" name becomes his "good" name. In the Bengali tradition, "good" names "represent dignified and enlightened qualities." In contrast, "pet" names "are frequently meaningless, deliberately silly, ironic, even onomatopoetic" (26). Thus, whereas pet names are used in the private family realm among family members and friends, good names are what the person is known by outside the family circle. Gogol's lost name is the *elliptical* space in which Gogol's good name and pet name have been conflated. That is to say, the name "Gogol" blurs that distinction between the intimate family network and the formal, outside-of-the-family social space. "Gogol" is a name that does not anchor its owner in either the Bengali cultural tradition or the American mainstream cultural tradition; it is neither an Indian Bengali name nor a common white American name. As a pet name it is a foreign name among Bengalis, and as a good name it is a foreign name among his classmates and coworkers. Unlike his parents, who by virtue of owning a house lay claim over a "small patch of America," Gogol is unable to lay claim over a place that physically exists. Early on, at the age of ten in the school fieldtrip to the graveyard, Gogol realizes that "his body will occupy no plot of earth, that no stone in this country [U.S.] will bear his name beyond life" because he will be cremated, not buried (69). As Himadri Lahiri underscores in "Individual-Family Interface in Jhumpa Lahiri's *The Namesake*," in the cemetery, Gogol is aware that he "does not have any ancestral history in the land that would connect him to any tradition in the national space; he is so different that his social and religious rite will be incompatible with those of the new country" (15). The act of burial and resting places per se as Anthony D. Smith underlines in *Myths and Memories of the Nation* (1999) are essential to ethnic groups in their acquired homelands. Smith explains that there is an intertwined relation between places and rooting memories in these places, and terms this process "territorialization of memories" (151). In *Chosen Peoples* (2003), Smith elaborates that

> after a few generations, the acquired homeland became "ancestral," the place of home and work, family and burial, for the community and its members. (148)

Gogol's realization that there will be no burial place for him that would mark his *place*, *home*, *work*, *family* suggests the *deterritorialization* of his memories and that they are not rooted in a singular place and are not associated with a chain of generations who belong to the same place. Gogol's inability to mark a place to locate himself can be interpreted in light of Michel de Certeau's theory about names, places, and meanings in *The Practice of Everyday Life* (1984). Certeau suggests two ways to locate oneself: naming a particular place and hence "giving it a meaning" (102); and telling stories about this place, so that they "tell us what one can do in it and make out of it" (122). Gogol's lost good name symbolically means his present and future dislocation. Unlike his parents who manage living outside their home country and own a "small patch of America," Gogol cannot imagine his place in America.

Throughout the novel, not only is there a direct correlation between Gogol's name and his problematic affiliation with a social/national space, but also between his name and his failed romantic relationship. In college, Gogol decides to use the good name his father has chosen for him as a formal name in school: Nikhil. It resembles the first name of Gogol (the writer), Nicholai and is not difficult to pronounce. In kindergarten, Gogol has insisted on keeping his name because this is what his parents, acquaintances, and friends call him. In college, Gogol decides to be Nikhil to symbolically exit from his parents' "traditional" life. When Gogol meets his first girlfriend—Ruth—as Nikhil, ironically, his only problem is that "he does not feel like Nikhil. Not yet" (105). Nevertheless, he is reluctant to introduce Ruth to his family, where "he is still Gogol" (115). Gogol as such occupies an interstitial space between being Gogol and being Nikhil: "'I'm

Nikhil now,' *Gogol* says" (119 emphasis added). This conflicted/confused *double consciousness* marks Gogol's temporary settlement in the dorm room, the shared apartment with college mates, as well as his apartment in New York.

Gogol roots himself, spatially in his dorm room as Nikhil and calls it *home*, a place that ironically he cannot occupy for long. After the dorm room, Gogol/Nikhil moves to an apartment shared with his friends. In this part of the novel, where his relationship with Ruth is about to be over, there is no description of the apartment or a reference to his name. Except for one sentence that describes "Gogol" spending his summer in Pemberton Road, the rest of this part uses the third person "he" to describe his break up with Ruth (119–120). The ambiguity of the space "he" occupies as well as the name associated with that space reveals a spatial non-national moment, emptied from any sense of affiliation. Similarly, when Gogol moves to New York to work for a midtown firm, he lives in a small building in a noisy street. The apartment does not reflect his settlement in any sense. There is a tea kettle that Gogol has never filled with water and a toaster he has never plugged in (126). Gogol's life in this apartment is one of anonymity; his walls are bare and his mailbox does not have his name on it. During the times when Gogol/Nikhil struggles with emotional frustration, he, like Ashima during pregnancy and in the hospital, conjures up the warmth of an imaginary home-space. Like the hospital and the apartment in Central Square, Gogol's dorm room and two apartments are social spaces that evoke a "missed" feeling of at-homeness.

In different parts of the novel, the spatial description of Gogol is in terms of either entering the "world," detaching himself from his parents' "world," or trying to create his own "world." As I have pointed out earlier, Ashima pities her son "entering the world so alone. So deprived" (25). Ashoke, however, happily writes to his family and relatives in India "Gogol enters the world" (29). When Gogol gets a job in New York, he stops going to his parents' house on weekends and being a part of "their world" (126). With Maxine (another girlfriend), Gogol is relieved to be "in her world" (150). Unlike Gogol, his sister, Sonia, is described as a "citizen of the world" (62) and as the "true American" (63) when in her "rice ceremony" she picks up the plates with soil and the dollar bill (symbols that foretell that she can become a real-estate agent or a businesswoman). Initially her name is Sonali, Gogol's sister is nicknamed and known as Sonia—a name that does not have as strong an ethnic resonance as do Gogol, Nikhil, Ashima, Ashoke, or Ganguli. This implies that mobility (the circulation of persons) and emphasis on ethnic identity are incongruent. In every attempt Gogol makes to discard his cultural heritage, he realizes that he is an outsider in the world of mainstream America that he strives to enter. For example, echoing the school fieldtrip to the graveyard, in the Ratliffs' [Maxine's family] graveyard, Gogol is aware that neither he nor his parents possess a similar place that will bear their names. While Maxine has a place to be buried (i.e., among her ancestors), Gogol cannot imagine his parents or himself buried in the United States. Gogol does not imagine his parents growing old in Pemberton Road, despite the fact that they actually do; he is unable to connect his parents past that is lingering in Calcutta to their present and future in Pemberton Road. Sonia does not encounter the same dilemma. Paradoxically, however, both Gogol and Sonia share the same "modern malaise called placelessness," to borrow Leonard Lutwack's term. Lutwack in *The Role of Place in Literature* (1994) holds that due to the growing economic process and the development of transportation and communication in the twentieth century, the importance of place has diminished and the importance of mobility has increased. As a result of increasing mobility and common technologies, "every part of the world is beginning to look like every other part" (183). Along the same lines, Edward Ralph defines placelessness as "the underlying attitude which does not acknowledge significance in places." It "[cuts] roots, [erodes] symbols, [and replaces] diversity with uniformity" (qtd. in Lutwack 183). Gogol looks for a place to be his, to mark it with his own existence, but in vain. Sonia, though, is not preoccupied with a spatial configuration of her own existence, does not really occupy a specific place, hence her flat characterization, so to speak, evokes a sense of "placelessness."

Thus far, social spaces in the novel—such as kitchen, hospital, apartments, neighborhoods—allegorize a spatial configuration of the non-national, which is tied to the experience of migration and creating a home-space in the United States. This raises a question about India as a home country: how does returning to the home country reconfigure the perception of India as a (non)-national space? When they move to Calcutta, during Ashoke's sabbatical year, the Gangulis stay in Ashima's mother's flat. The sense of collectivity and family ties is stronger in India. Thus, instead of having an apartment of their own, the Gangulis stay with different members of their family. Therefore, "[e]very few weeks there is a different bed to sleep in, another family to live with, a new schedule to learn" (83). Despite the fact that Ashima and Ashoke are among their large family in India, moving between houses implies a sense of unsettlement and becomes as well part of the *parenthetical* status of being foreigners even in India. In their trip to Delhi and Agra, the Gangulis "are tourists, staying at a hotel with a swimming pool, sipping bottled water, eating in restaurants with forks and spoon, paying by credit cards. Ashima and Ashoke speak in broken Hindi" and Gogol and Sonia speak only in English. By the same token, the Gangulis visit touristic places and take pictures to commemorate their visit (84–85). They are received and welcomed as visiting rather than resident family members. For example, Ashima does not set a foot in the kitchen for eight months. Although the Gangulis do not face the same dilemma of un-belonging that the Egyptian returnee—Hani—encounters in *The Time between Places*, the description of their time in India echoes the demarcation made in *He Had Dreamed of Returning* between those who have migrated and those who have never left and stayed put. Migration problematizes the correlation between filiation and national culture, and unfolds the migrants' occupation of a third space in-between the national and the foreign.

The non-national spaces between the national and the foreign are moments of spatial and/or emotional solitude that underscore the binary of national-foreign. In the kitchen scene, approximating an Indian snack, Ashima's frustration provokes this sense of occupying a non-national space as I have clarified earlier. Gogol, although born and raised in the United States, occupies a similar space when he, paradoxically, embraces his Bengali heritage, especially after his father's sudden death. Gogol revives his affiliation with his parents' Bengali tradition in two ways: emulating his father's rituals and marrying a wife with a Bengali lineage.

In the apartment that his father has rented during his temporary stay in Cleveland, Ohio, Gogol imagines his father occupying the space of his apartment before he drives to the emergency room. Gogol pictures his deceased father moving around the apartment in the same way he has known him. Imagining the daily routine of his father echoes the scene where Ashima imagines having an Indian snack or giving birth in Calcutta. Ashima's frustration resonates with Gogol's grief; they both share feelings of alienation and homesickness. That is to say, Ashoke's daily routine is precisely what Gogol misses and provokes feelings of nostalgia and vulnerability. This echoes Ashima's feelings of vulnerability and nostalgia at the opening of the novel. To "compensate," Gogol, in addition to performing some of his father's tasks (such as paying the bills for his mother and shoveling the snow), decides to marry an American-born Bengali wife: Moushumi. Every time Gogol sees Moushumi, he recalls an image/incident from their childhood. However, although Gogol looks for his Bengali heritage in his marriage to Moushumi, what their relationship emphasizes is the fact that they both occupy a third space that does not fit in the either-or categories of national or foreign. Like Gogol's name that is neither Bengali nor mainstream American, Moushumi immerses "herself in a third language [French], a third culture.... It was easier to turn her back on the two countries [India-U.S.] that could claim her in favor of one that had no claim whatsoever" (214). Gogol admires Moushumi for her capacity to have a "separate life" in a different country (neither India nor the United States) (233); and Moushumi admires Gogol for changing his name to Nikhil. That is to say, what they share is precisely what they do not want to acknowledge sharing: their reluctance to adopt their Bengali heritage. This ambivalence, paradoxically, has led to their marriage to fulfill "a collective, deep-seated desire—*because they're both Bengali*" (224 emphasis added). Are they? Nikhil and Moushumi's apartment problematizes the notion of home as a sign of national culture. Though small like the Gangulis' first apartment in Central Square, it is "luxurious" and in the heart of New York city. It is artistic rather than practical: built-in mahogany bookcases, stainless-steel appliances, marble floors and walls in the bathroom, and a Juliet balcony off the bedroom (228). They furnish it with brand-name imitations. Nikhil and Moushumi's lifestyle is like their apartment: eclectic, with no glimpse of approximating Bengali tradition. Instead of replicating a specific national culture, Nikhil and Moushumi's apartment signifies an ambiguous non-national space.

The closing of the story blurs the national and the foreign in two contradictory ways when Ashima decides to sell their house in Pemberton Road and when Nikhil decides to give a new turn in his life in which he reembraces his forsaken name (Gogol). In a way, echoing Kaldas's story *A Conversation* in *The Time between Places*, India is imagined from Ashima's perspective as a prospective home. The *dialogic* spatial imagination of the United States and Egypt in *A Conversation* recurs in *The Namesake.* On the one hand, the home in Pemberton is the one "she has created" though "she still does not feel fully at home within [its] walls" (280). On the other hand, Ashima is returning to Calcutta "the city that was once home and is now in its own way foreign" (278). However, unlike *A Conversation*, Ashima does not perceive home country and host country in a binary opposition but rather encompasses both the United States and India as *home(s)* when she decides to spend six months in India and six months in the United States. Although the binary between home country and host country in *A Conversation* has unfolded the couple's lack of full belonging to either the United States or Egypt, Ashima's decision to be a part-time resident in both India and the United States ultimately implies lack of full belonging in either place. Ashima "will be without borders, without a home of her own, a resident everywhere and nowhere" (276). Ashima's "placelessness" dismantles a spatial imagination of the nation based on the national-foreign dichotomy. Whereas Ashima goes beyond the boundaries of a singular sense of belonging, Nikhil reroots himself in his old forsaken self: Gogol. In the last scene of the novel, Nikhil takes refuge in his room and "starts to read" *The Overcoat* by Nicholai Gogol (291). Imagining his mother interrupting him, Gogol turns over a small corner of a page "to mark his place" (290). Gogol/Nikhil strives to root himself. Ironically, where he decides to root himself is the same third space marked by his name that is neither Indian nor mainstream American. That is to say, both Gogol's and Ashima's perceptions of *home* are two faces for the dynamics between residing *anywhere* and *nowhere*.

In conclusion, *The Time between Places* and *The Namesake*, though they portray the pain and hardships of migrating and the ups and downs of pursuing better living conditions in the United States, they depoliticize the narratives. To clarify, the time frame for both texts is the 1960s. Although there were crucial political events taking place in the 1960s (such as the Vietnam War, the Cold War, and the Civil Rights movements), neither text highlights any political realities that the characters might have encountered. On the contrary, in one of the early scenes in *The Namesake* when Ashima decides to be independent and venture exploring Central Square on her own accompanied by her baby boy Gogol, the scene reveals passersby who are curious about her son and are warm to her and her son. In the same sense, Kaldas, although she accentuates the impact of Nasser's rule on the lives of different families in Egypt, there is no mention whatsoever of any political realities in the United States that might have influenced the characters' lives. The characters' political disengagement is a form of un-belonging, which I interpret as non-national. In "He Had Dreamed of Returning," Kaldas describes how Hani's family follow up with the news of the war in Egypt in 1973, then continue "with their lives the next morning" (151). Likewise, Lahiri describes how Ashoke and Ashima, along with other Bengali/Indian families, spend hours arguing about the politics of America, "a country in which none of them is eligible to vote" (38). That is to say, home in *The Time between Places* and *The Namesake* reveal that home and nation are separate spheres.

CHAPTER 4

Transnational Allegories and the Non-national Subject in *The Agüero Sisters* and *The Night Counter*

The stories of migration in the narratives that I have discussed in the previous three chapters underline three aspects of the non-national in contemporary narratives by American women of color. These aspects have revisited notions of unified national consciousness, linear national time, and a homogeneous sense of belonging to a singular national space. My analysis in the earlier chapters reveals a different sense of the nation that is formed beyond geographical boundaries and mediated by locating home(land) cognitively and spatially rather than solely perceived as a place of roots and belonging. That is to say, the non-national disrupts the idea of successive generations handing down the nation as if it is an invariant substance, inherited from one generation to the next. In addition, the non-national reconfigures the nation spatially as an open, un-confining space. This provokes a question about how to interpret the story of the nation in migrant narratives if the perception of the nation is not confined to national territories or geographical boundaries. In this chapter I analyze *The Night Counter* by Alia Yunis vis-à-vis *The Agüero Sisters* by Cristina García to suggest reading these texts as national and transnational allegories and to pose a hypothesis about the non-national subject as a possible space for allegory.

The question about allegorizing the nation conjures up multiple implications for allegory. Walter Benjamin's critical study of German plays of mourning, the *Trauerspiels*, in *The Origin of German Tragic Drama* (1985), restores the literary significance of allegory, which was replaced by symbolism especially by the Romantics. Benjamin elaborates that allegory originates in melancholy and mourning induced by historical crises. Allegorists, as Benjamin explains, see death and decay in both nature and history. They go beyond the beauty of nature—unlike the Romantics who celebrate nature—seek its contradictions, and view death as an essential part of nature (178). Benjamin underscores that allegorists seek "the particular from the general" and "the particular serves only as an instance or example of the general" (161). Katherine Sugg, deploying Benjamin's theory, reemphasizes in *Gender and Allegory in Transamerican Fiction and Performance* (2008) that allegory is closely connected to historical crises. She elaborates that while postmodernity is the historical crises of late capitalism, post-9/11 is the historical crises of contemporary times (33). Craig Owens's analysis of allegory as a narrative form in "The Allegorical Impulse: Toward a Theory of Postmodernism" (1980) also aligns allegory with postmodernism in the sense that "[t]he allegorical work is synthetic, it crosses aesthetic boundaries," producing a "confusion of genres" (1055).

Building on the interlocked relation between the ***particular*** and the ***general***, Gayatri Chakravorty Spivak and Frederic Jameson focus on the political dimension that allegories in fiction can convey. On the one hand, Spivak analyzes *Jane Eyre* in "Three Women's Texts and a Critique of Imperialism" (1985) and holds that the novel can be read as "an allegory of the general epistemic violence of imperialism" (251). On the other hand, Jameson, in "Third World Literature in the Era of Multinational Capitalism" (1986), analyzes novels in third world literature (exemplified by Chinese and Cuban novels) to argue that these novels are political, specifically in their portrayal of the battle against the imposition of capitalism on their nations and hence should be read as "national allegories." Allegory in Jameson's essay is a "signifying process, which might only be set in motion and complexified were we willing to entertain the more alarming notion that such equivalences [figures and personifications] are themselves in constant change and transformation at each perpetual present of the text" (73). Focusing on third world literature, Jameson holds, "the story of the private individual destiny is always an allegory of the embattled situation of the public third-world culture and society." Thus, all third world texts "necessarily project a political dimension in the form of national allegory" (69). In Jameson's essay, allegory addresses the problem of "narrative closure" (76) and the absence of a specific form of closure in third world literatures, produced after independence. This problem is evoked by the political dilemma of realizing that postindependence is not necessarily a better era, economically or politically, than the colonial era. Thus the individual subject is unable to grasp his/her whole situation. He suggests that reading allegorically generates "a range of distinct meanings or messages, simultaneously, as the allegorical tenor and vehicle change places" (74). Building on Jameson, I argue that *The Night Counter* and *The Agüero Sisters* are allegories in the sense that they generate "a range of distinct meanings or messages, simultaneously." In each text, the individuals' stories are open stories and are tied to a "national situation": post-9/11 in *The Night Counter*, and post-1959 socialist Cuba in *The Agüero Sisters.*

Although Jameson's concept of national allegory is a useful analytical tool, it is also controversial. A case in point is Aijaz Ahmed's criticism in "Jameson's Rhetoric of Otherness and the 'National Allegory'" (1987). Ahmed criticizes the definitional limitations of "national allegory," especially the first-world—third-world binary on which Jameson's argument is founded. Ahmed holds that minorities in the United States form "pockets" of the third world. Ahmed, in addition, replaces the notion of the "nation" with "the larger, less restrictive idea of 'collectivity'" (88). In light of Ahmed's argument, "allegorizing" individual experience is not necessarily nationalistic (referring to the experience of colonialism and imperialism and the struggle for liberation), and is not specific to first or third world literature. Ian Buchanan, in "National Allegory Today: A Return to

Jameson" (2005), takes the "national" in Jameson's essay to mean "national situation" rather than "nationalist ideology" (174). In light of Buchanan's interpretation, I argue that Jameson's concept does not account for a complex, problematic relationship between the individual and the nation where the individual is dis-placed from one nation, relocates in another, and adapts to/adopts the new nation as an alternative homeland. In other words, Jameson's concept of "national allegory" does not take into account the question of assimilation or integration for migrants, hence does not include the non-national subject or problematic "national situations." Similarly, although Ahmed replaces the notion of the nation with "collectivity," Ahmed does not count for conflicted collectivities. For instance, as I discuss in *The Night Counter*, members of the same family show different alliances in the aftermath of 9/11. Some family members emphasize their religious identity and even become conservative Muslims; while other members assist the FBI agents interrogating terrorist suspects. Furthermore, "time-space" compression that characterizes the formation of "imagined (transnational) communities" unsettles a homogeneous perception of the nation/collectivities. Therefore, I argue that stories in *The Night Counter* and *The Agüero Sisters* are open-ended, because they are not situated within a singular nation but rather reveal intertwined political and economic histories between the United States and the Arab region, as in *The Night Counter*, or the United States and Cuba, as in *The Agüero Sisters*, that is to say, the nation in these narratives is not limited by geographical boundaries.

My argument here resonates with Jean Franco's proposition in "The Nation as Imagined Community" (1997) about "whether the term 'national allegory' can be any longer usefully applied to a literature in which 'nation' is either a contested term or…a mere reminder of a vanished body" (131). Imre Szeman points out in "Who's Afraid of National Allegory?" (2005) that the nation in Jameson is conflated with the political, whereas it should be put into question (198). That is to say, Jameson's concept of "national allegory" does not consider transnational relationships, specifically formed as a result of economic unevenness and asymmetries that create third world economic conditions within the United States. As Buchanan says, "the Third World loses all geographic specificity and becomes instead a term designating something like a new class or indeed caste system within the First World itself" (178). Building on Buchanan, I resituate the third world within the United States, especially in my discussion of economic unevenness, in *The Night Counter*, between Arab-Gulf states that produce oil and invest in the United States, and Arab migrants who live at the margins of that economic luxury. Elaborating on Jameson, I argue that *The Night Counter* and *The Agüero Sisters* are transnational allegories, in the sense that they are political critiques of monolithicizing Arab Americans and Cuban-Americans in the United States, and of economic disparities caused by global capitalism. I also extend Jameson's concept of "national allegory" and argue that the non-national subject can be read allegorically, generating multiple meanings, and specifically in *The Night Counter*, it can be read as a critique of systematic erasures of the ethnic subject in the United States.

The Night Counter depicts three generations of the same family, born and raised in the United States; yet, despite their long presence in the United States, members of this family struggle to be acknowledged as fellow American citizens (rather than potential terrorists), especially after 9–11. *The Agüero Sisters* portrays three generations of Cuban and Cuban-American families who, upon moving to the United States, are accepted within mainstream society—unlike Arab Americans—largely because Cuban exiles oppose the Castro regime, which several American administrations have condemned as a threat to the United States. In both texts, families are divided along the lines of political and economic interests that can facilitate and/or hinder their integration as American citizens in the United States. In Yunis's text, family members are divided because of the Patriot Act and Homeland Security measures that the American administration has taken against individuals of special interests (i.e., suspected terrorists). Thus, while some family members are under surveillance by the FBI, other members of the same family are helping the FBI to investigate other Arabs and Arab Americans in the United States. In *The Agüero Sisters*, members of the family are dispersed between Cuba and the United States because of their belief in either the capitalist world of the United States or the communist revolution in Cuba.

I argue that while *The Night Counter* and *The Agüero Sisters* can be interpreted as national allegories; they also problematize the notion of the nation as an economic and political territory. That is to say, my analysis of *The Night Counter* and *The Agüero Sisters* places the notion of the nation in Jameson's concept of "national allegory" into question. To that end I will start with an analysis of *The Agüero Sisters*, which can be read as a national allegory in the Jamesonian sense, but with a glimpse of some aspects of the non-national, before analyzing *The Night Counter* as an example of what I term "trans-national" allegory.

The Agüero Sisters

Cuban literature in the United States dates back to the nineteenth century when Cuba was a Spanish colony and Cuban expatriates who fled to the United States produced literature addressing the fight for Cuba's independence during the Spanish-Cuban-American war. In his survey of the history of Cuban American literature, titled "History and Development of Cuban American Literature: A Survey" (1993), Rodolfo J. Cortina points out that different groups of Cuban writers have come to the United States for different political and economic reasons. Along these lines, Cortina draws a distinction between Cuban authors who write in Spanish about Cuba in the United States, on the one hand, and the Cuban American authors who use English as their mediating language, on the other hand. For the nineteenth-century expatriate authors, the Cuban revolution was the answer and hope for an independent Cuba, whereas for later groups of writers who have come to the United States with their parents in the 1950s, or were born in the United States, the revolution was a devastating economic and political turnover. Yet, as Cortina elaborates, there is a third group of writers who aim their criticism at the United States, specifically its economic exploitation of migrant labor, as well as at the Cuban exiles in the United States. Similarly, Jorge Duany, in "Neither Golden Exile nor Dirty Worm: Ethnic Identity in Recent Cuban-American Novel" (1993), holds that Cuban-American writers who have emerged in the 1980s and afterward transcend, in their narratives, the stereotypical perception of Cuban exiles in the United States as either political refugees (forming successful economic enclaves), or as *marielitor* (lower class and impoverished migrants). This more recent group of writers address social and cultural problems among Cuban migrants in the United States. In her study, *Cuban-American Literature of Exile: From Person to Persona* (1998), Isabel Alvarez-Borland designates this group of authors, who come to the United States with their parents as infants during the Cuban revolution as "Cuban-American ethnic writers" (7). She elaborates that "Cuban-American ethnic" share some of these aspects in their narratives: writing to both Cuban exiles and non-Cuban readers in the United States, emphasizing Cuban heritage, and constructing an American-Cuban cultural identity. One can situate Cristina Garcia's works within this literary, historical framework.

Cristina García is a renowned Cuban-American writer mostly known for her first novel, *Dreaming in Cuban* (1992), which is more critically acclaimed than her second novel, *The Agüero Sisters* (1997). Among her other works are *Monkey Hunting* (2007), *The Lady Matador Hotel* (2010), and her latest, *King of Cuba* (2013). Most scholarly work of García addresses the correlation between family histories and the history of Cuba since the early twentieth century, the tension between Cuban exiles in Florida and Cubans (provoked by the 1959 revolution), the diversity of Cubans and Cuban-Americans, and the influence of Afro-Cuban traditions on García's novels.[1] In his survey of Cuban-American literary history—*One Island, Many Voices: Conversations with Cuban-American Writers* (2008)—Eduardo R. Del Rio proposes that Cuban-American literature since the 1980s not only continues to describe the conflict between Cubans and Cuban-Americans but also depicts the US economic system as an exploiter of Cuban workers (5). Along the same lines, Teresa Derrickson in her critical study, *Politicizing Globalization* (2002), suggests that *The Agüero Sisters* not only depicts the historically contentious relationship between Cuba and America, but also "makes clear that the female body is a site where state and global politics play themselves out" (157). I expand on these critical studies by situating *The Agüero Sisters* within and against Jameson's theoretical concept of national allegory.

As I have mentioned earlier, *The Agüero Sisters* portrays three generations of Cuban and Cuban-American families. There are three story lines in this novel: the story of Ignacio and Blanca Agüero; the story of Constancia and her family who have migrated to the United States after the 1959 revolution in Cuba; and the story of her younger sister, Reina, and her daughter, Dulce, who stayed in Cuba, yet eventually leave for the United States. The incidents of the novel take place in both Cuba and the United States and are narrated from the perspectives of Ignacio Agüero, Constancia and her children Isabel and Silvestre, Reina and her daughter Dulce. The framing story is Ignacio's diary that recounts his Spanish father's quest to the New World—Cuba in the late nineteenth century—and the circumstances of his wife's—Blanca—murder. Ignacio's life—from the early to the mid-twentieth century—as narrated in his diary, alternates with Constancia's and Reina's stories in Cuba and the United States in the early 1990s. These alternating perspectives reveal the political and economic changes that have taken place within Cuba and also reveal the American involvement.

Ignacio and Blanca are naturalists who share an interest in Cuba's flora and fauna. Constancia tells her story in New York, then in Miami; and Reina describes her life in Havana then in Miami. In addition, there are substories—Silvestre's, Isabel's, and Dulce's—that resonate with these three main story lines. At the center of the novel is Blanca's enigmatic death. The prologue alludes to her death and to the fact that Ignacio lies about her death. In the very last page of the novel, the mystery of Blanca's death is explained and Ignacio is revealed as her killer. Ignacio's diary, as well as his academic legacy, causes tension between the two sisters, Constancia and Reina. Constancia is jealous that Reina has more access to their father's books by virtue of staying in Cuba, and Reina is jealous that Constancia knows more about their mother and the circumstances of her untimely death at the age of 35. The death of Blanca that is mentioned without details in the prologue is narrated in fragments throughout the novel.

The novel draws analogies between Blanca and Cuba's birds, botany, indigenous creatures, and natural resources throughout. When Ignacio first sees Blanca, he compares her to a bird: "slight" and "delicately boned as certain birds," with "distinctive camouflages of her subjects" and "versatile physiognomies" (183). Her green eyes take a yellowish tint when she works with sulfur, and when she works with phosphorous, she vibrates with "its unearthly glow." Like "lead," Blanca appears "heavy and malleable and gray" (182–183). Blanca appeals to and intrigues a naturalist like Ignacio, and she makes him curious to explore his "object" of study. Ignacio's character echoes the Spanish naturalists and their expeditions to the Caribbean Islands and Latin America that started in the mid-eighteenth century, which used scientific knowledge to impose their dominance over these territories and peoples. As Stuart George McCook puts it in *States of Nature* (2002), "The Spanish imperial government...used science to nationalize nature, to extend state power over the natural world" (11). Ignacio's interest in Blanca can be interpreted in light of his interests as a naturalist during the colonial era in Cuba. Blanca's name means white or blank, which metaphorically suggests Blanca is a new horizon for Ignacio to pursue. That is to say, Blanca and Cuba are both open for colonial exploration as well as for exploitation. In the prologue, Ignacio says: "Long ago, Cuba had been a naturalist's dream" (4). Later in his diaries in a moment of reflection on Cuba's nature, Ignacio writes:

> I thought of what the *first explorers* must have felt at *the sight of a new horizon*, at *the roar of possibilities* in their heads. How they imagined the vast riches that awaited them, all there for the *taking* with a *musket* and a *strong pair of hands*. (92 emphasis added)

The passage expresses the novelty, the abundance of possibilities and potential wealth the land offers, on the one hand, as well as the "right" to take them by force, "with a musket and a strong pair of hands," on the other hand. One can draw a parallel between Ignacio's fascination with and desire to capture this new horizon and his fascination with Blanca and his desire to control her. Ignacio's failure to control Blanca frustrates him and leads to her murder. Her murder is described as a mistake, when Ignacio tries to shoot "the most extraordinary bird" hovering above Blanca's head (299). Ignacio describes his shooting of Blanca as "a necessity of nature" where the bird/Blanca itself/herself invited him to capture it/her (299). The "untimely" death of Blanca is a "sorrowful" incident, caused by Ignacio that shatters the unity between Blanca and Cuba's indigenous birds. Ignacio's books/articles about Cuban plants and creatures include: *Cuba's Dying Birds* (155); *Cuba: Flora and Fauna*; *A Naturalist's Guide to the Pearl of the Antilles* (11); "The Lost Reptiles of Cuba" (185). These titles stand for a sense of nationhood formed, not through specific political/ideological views, but by virtue of claiming knowledge and hence power over Cuba's flora and fauna.

The correlation between Cuba's indigenous birds and the nation in *The Agüero Sisters* is also present in the details Ignacio gives about the day of his own birth (1904, two years after the independence of Cuba in 1902). The day Ignacio is born is jinxed by the appearance of an ill-omen, an owl that plucks the placenta of Ignacio's mother and spreads birthing blood over the crowd gathering for the presidential inauguration (29). Despite the implications of the ill-omen owl, birth blood can also mean the birth of the new Republic after independence and the beginning of a new era. However, it is not a very promising birth, because despite Cuba's independence from the Spanish, the Platt Amendment signed in 1901 between the United States and Cuba (a Spanish colony at the time) was still in effect. This Amendment defined the geographical boundaries of Cuba, facilitated leasing the Guantanamo Bay as a naval base for the United States, and controlled Cuban trade (specially opening new markets in Cuba for US products in exchange for exporting Cuban sugar to the United States). This means that Cuba's *national project* of independence was curtailed from the very beginning. Cuba's independence ended Spanish colonialism, but not American imperialism. Despite the dominance of the tobacco industry in Cuba in the early twentieth century, the sugar boom took over and many cigar workers lost their jobs "to the modern cigar-rolling machines from America" (113). Cuba's political independence in 1902 opened more Cuban markets to the expansion of American capitalism. This direct correlation between the jinxed birth of Ignacio and the political independence of Cuba—but not its economic sovereignty due to the Platt Amendment—can be read as a national allegory in the Jamesonian sense: "the story of the private individual destiny is always an allegory of the embattled situation of the public third-world culture and society." The embattled situation in Cuba was the false hope of independence that provoked the 1959 revolution and a new national project, namely, the transformation from capitalism and the free market to socialism and state ownership.

There is a hint in the novel that the Agüero house is a big house owned by the family. Yet after the revolution its eight rooms were redistributed among many families, indicating that the new regime after the revolution in Cuba has affected negatively the economic status of the Agüero family. That is to say, the economic changes that have taken place in Cuba after the revolution were not in the best interest of the Agüero family. Thus, Constancia leaves for the United States, while Reina stays on. Whereas Reina is tolerant of the communist regime, Constancia cannot live under its grip and hopes it will end. The novel as such depicts two different Cuban nationalisms from the perspectives of Cubans and from the perspectives of Cuban-Americans. For example, Reina works under the communist regime in Cuba and does not mind it, while Constancia despises it and lives in self-exile in the United States.

Though Reina and Constancia disagree on what is good for the nation, they agree on devalorizing patriotism and nationalism. For example, Constancia criticizes her husband's enthusiasm for the exile group, La Brigada Caimón, that aims at deposing the communist regime in Cuba. She says: "Men always confuse *patriotism* with self-love! . . . war should be strictly personal, like philosophy or sexual preference" (77 emphasis added). By the same token, Reina reiterates that "*patriotism* is the least discerning of passions" and is sarcastic about the kind of Cuban nationalism propagated over exile stations in Miami: "[it is] *parading nationalism*, like a bunch of roosters in the make" (196 emphasis added). While Constancia speaks harshly about Cuban patriotism among the Cuban exiles, she is considered not Cuban enough by other Cuban exiles because she once voted for a Democratic president. Cuban exiles are perceived by their compatriots in Cuba as either traitors because they have left the country, or the real heroes because they have made it to the United States.

In *The Agüero Sisters*, there are two stories/perspectives of the nation—mediated from inside Cuba and from the United States. Both stories are initially set in opposition within the framework of the eternal conflict between communism and capitalism. The clash between conservative and liberal economic systems in Cuba, together with the failure of the Cuban regime to deliver its promises, leads me to read this novel as a "national allegory." Dulce's unconcealed frustration because of the living conditions in Cuba "unravel[s] the revolution" (68). Instead of equal jobs and equal income, there are no decent jobs and not enough money. Cubans are "sick of picking potatoes and building dormitories, only to find no meaningful work in the careers [they have] trained for. . . . Sick of having nothing to do, period" (52). In addition, Dulce compares Cuba to "an evil stepmother, abusive and unrewarding of effort. More, more, and more for more nothing" (52–53). Reina, later in the novel, echoes Dulce, when she laments the lost beauty of Cuba and compares the times she used to walk with her mother by the Almendares River with what she and her daughter Dulce now see: "filthy, shimmering with mosquitoes and algae, the trail clotted with garbage and rotting fruit" (99). This description of the Almendares River stands in stark contrast to Ignacio's description of Cuba's exhilarating wildlife. In addition, Dulce's description of the social/economic conditions overshadows the gains and benefits that her mother's generation reaped after the revolution, especially its women. Unlike her mother who is denied a job because she is pregnant, Reina has a government job as an electrician. Reina has her daughter out of wedlock with José Luis Fuerte, one of the revolution's heroes. Never married, Reina is a long-term mistress of a married governmental official, Pepín Beltrán. All this shows that Cuban women after the revolution had a sense of independence and equality with their male peers. These benefits of the revolution are hardly highlighted in the novel. Except when Reina criticizes the exile stations in Miami for telling "outrageous lies on the air" about the Cuban regime and repudiating even the achievements of the revolution; these accomplishments are subsumed by Dulce's narrative voice and her jobs as a teacher and a prostitute to make ends meet.

In contrast, Constancia is portrayed in her first appearance in the novel in the heart of New York, making 1200 dollars in sales in the first half hour of the day (20). Since the circulation of money does not see color/language/geographical boundaries, Constancia's business booms even more when she founds her own factory of beauty products—Cuerpo de Cuba (body of Cuba)—in South Miami. Although the circulation of capital may cross color, language, and national boundaries, ironically it is these three factors that help Constancia's sales to rise. Constancia makes 1200 dollars in half an hour by virtue of the way she looks and talks:

> her skin is *soft* and *white*. Her dark hair is arranged in a *French bun*, and nails are lacquered to match her carnelian lips.... she completes every ensemble with a short strand of pearls. Her *foreign accent* and *precise* manner *intimidate* clients into buying whatever she suggests. (20 emphasis added)

The detailed description of Constancia's appearance reflects contradictory implications. Constancia's accent does not detract from her appeal to her customers. On the contrary, Constancia's accent renders her products exotic, and hence attractive by virtue of her difference. To recall Rutherford's *Identity: Community, Culture, Difference*, from the first chapter, "capital has fallen in love with difference" (11). Constancia's intriguing *difference*, though marks her as *ethnic*, her skin color renders her *ethnic* but *not quite*. That is to say, what helps Constancia to be appealing is not only her confidence as a seller, but rather her "white" skin and her fashionable attire that intrigue some women to emulate her, and attract men as well.

Constancia manipulates "Cuba" to expand her business. Echoing the titles of her father's books in which Cuba is at the center, Constancia's beauty products include: "Ojos de Cuba" (an eye repair cream); "Pies de Cuba" (foot bath); "Cuello de Cuba" (neck cream); "Senos de Cuba" (breast lotion); "Codos de Cuba" (elbow moisturizer); "Muslos de Cuba" (thigh cream) (129, 131). These names echo Ignacio's book titles in the sense that Constancia *nationalizes* Cuba through her beauty products, which create "an imagined community" formed around these products in New York and Miami. Not only does Constancia *nationalize* Cuba but also *engenders* it by correlating Cuba to a woman's body: "Cuerpo de Cuba." Nevertheless, metaphorically fragmenting the Cuban body into eyes, foot, neck, breast, elbow, and thighs renders "Cuerpo de Cuba" a trope for *reifying* "Cuba" as an object circulating to facilitate marketing beauty products. In addition, Constancia's line of perfumes is "Flor del Destierro" (flower of exile) publicized by echoing some verses from José Martí's poem "Flores del Destierro":[2]

> Que en blanca fuente una niña cara,
> Flor del destierro, candida me brinda
> Naranja es, y vino de naranjo. (271)
> [In a white fountain, a child's face,
> An exiled flower, it toasts to me simply,
> 'It is orange and orange wine']

Both Ignacio's books and Constancia's beauty products produce Cuba as a fragmented woman/national body. That is to say, the fragmented woman's body becomes what Kristeva calls, a "symbolic denominator" for Cuba. Analyzing the intertwined relation between women's bodies and the national, Kristeva in "Women's Time" defines the "symbolic denominator" as "the cultural . . . memory forged by the interweaving of history and geography" (188). Hence, many Cuban women admit that "they feel more *Cubana* after using her [Constancia's] products." "Cuerpo de Cuba" products make Cuban women recall "long-forgotten details of their childhood in Sagua la Grande, Remedios, Media Luna, or Santa Cruz del sur. . . . Politics may have betrayed Constancia's customers, geography overlooked them, but *Cuerpo de Cuba products still manage to touch the pink roots of their sadness*" (132 emphasis added). Constancia's products bring specific memories to their users of specific provinces in Cuba. What Constancia is successful at, here, is evoking a feeling of national affiliation through a circulation of nostalgic *affect*. Bringing together beauty products and national sentiments has double meanings. First, it shows Constancia manipulating the Cuban exiles' feelings of nostalgia and as a skillful seller, she addresses those feelings. Second, what Constancia is trading is not only beauty products but also "Cuba" as a national *affect*. This interconnection between circulation of money (through the circulation of beauty products) and targeting Cuban exiles' nostalgia and homesickness resonates with Sara Ahmed's notion of "affective economy" in *The Cultural Politics of Emotion* (2004). Ahmed, borrowing from Marx's theory of the movement of commodities and money, draws an analogy between emotions and capital. She says: "emotions work as a form of capital: affect . . . is produced only as an effect of its circulation" (120). Thus, "Cuerpo de Cuba" beauty line is not only a purveyor of beauty and femininity, but also of nostalgia and national affiliations.

The circulation of a nostalgic-national affect does not lead to a homogeneous collectivity. As I have mentioned earlier, Constancia feels that she does not fit among the Cuban exiles in Miami and despite the fact that Constancia's business booms in Miami by virtue of her compatriot consumers, she does not feel comfortable with their excessive feelings of nostalgia. To Constancia, Miami is "disquieting" and its air is "thickly charged with *expiring dreams*" (46 emphasis added). When Reina joins Constancia in Miami, Reina thinks of Constancia in the same exact way Constancia thinks of the Cuban exiles in Miami: "[Constancia] sounds like the past.

A flash-frozen language, replete with outmoded words and fifties expressions" (236). On the one hand, among the Cuban exiles Constancia feels a "foreigner" (45). On the other hand, Reina's perception of Constancia is as a Cuban exile haunted by the past. Constancia's character can be read as an allegory of "imagined (transnational) communities" mediating between Cuba and the United States rather than an allegory of a single nation.

The Manichean representation of the communist and capitalist worlds in the novel highlights dire conditions in Cuba (and the failure of its economic systems) and the feverish pursuit of money and wealth in the United States. In Cuba, people appreciate a roll of toilet paper or a bar of soap over leaflets inciting an uprising (239). In the United States, people are too busy accumulating money and wealth (199). The economic and political conflicts between both worlds is metaphorically represented in Silvestre's and Dulce's bodies.

Silvestre is a deaf young man who sorts out news for a news magazine in New York. Silvestre is Constancia's eldest child from her first marriage to Gonzalo Cruz (who is also the brother of her second husband, Heberto). Silvestre's considers his deafness as a good thing for his job because it "filters out distractions" (23). In the second chapter of Constancia's story, Silvestre's disability is described from Constancia's perspective as "[a]nother causality of that *dichosa* [prosperous] revolution" (44). He has been the victim of the political conflict between the United States and Russia. Constancia, hearing a rumor that the Cuban government intended to send Cuban children to boarding schools in Ukraine, decides to send Silvestre to an orphanage that hosted other Latino/a children in Denver, Colorado. Silvestre, however, could not stand the severe cold, got a fever, and lost his hearing senses (82). Silvestre is a victim of the conflict between communism in Russia and capitalism in the United States, each superpower wanting to exploit other nations to serve its interests. Before the 1959 revolution, the tobacco industry waned and was replaced by the sugar industry, which the United States encouraged. The communist revolution tried to manage its own economy but to no avail. Silvestre's disability is an allegorical critique of both capitalism and communism. Is it the revolution that caused Silvestre's deafness or the weather in Colorado? Imitating his mother's voice, ironically, Silvestre enunciates: "*Be a little man and don't cry, Silvestre. This is* [leaving Cuba for the orphanage in Colorado] *much better than becoming a communist*" (82). His mockery renders her decision meaningless. In allegorical terms, Silvestre's story can stand for the *embattled situation* in Cuba and the failure of the revolution to fulfill all its promises, but Silvestre does not live or grow up in Cuba; he becomes deaf and grows up in the United States. Silvestre lives at the heart of one of the most powerful capitalist centers in the world. If we read his deafness allegorically, working on *sorting* news, vis-à-vis Constancia *sorting* what she sells, and Reina's comment about people in the United States, "too busy *sorting* through the hysteria of what to purchase next" (199 emphasis added), one can read Silvestre and Constancia as two faces of an obliviousness to capitalist exploitation. Silvestre is literally deaf within the system he works for and Constancia's booming sales depersonalizes labor by concealing the abuses of a capitalist economic system.

The foil for Silvestre's and Constancia's bodies is Dulce's body. The chapters dedicated to Dulce's story are narrated from the first-person point of view, like Ignacio's chapters, which are mainly his diary, and unlike the rest of the chapters, which are narrated from the third-person perspective. Dulce's story can be read as the counter-perspective for Ignacio's story. Unlike Ignacio's fascination with Cuba's wildlife that is every "naturalist's dream," as I have previously pointed out, Dulce is frustrated with the deplorable living conditions in Cuba. When Dulce thinks of a dream place, she thinks of New York: cold weather, skating, fur clothes, and frozen lakes, and daydreams of herself skating in circles. Dulce does not see any promise in Cuba and "the future is frozen" (54). Instead of details about Cuba's birds, turtles, fauna, lizards, owls, Iguanas, Dulce describes the economic hardships that hit Cuba and have made many Cubans desperate for "leaving and dollars" (54). A volleyball coach in José Martí High school, to make ends meet, Dulce wanders Cuban touristic areas to earn "pocket money"

(52). Dulce's body, hence, becomes an economic resource and this parallels Constancia's use of "Cuba" as a body to sell her beauty products. While Constancia fragments "Cuba" into eyes/hands/feet/face rejuvenating products, Dulce desensitizes her body. When Dulce moves to Madrid with her 64-year-old boyfriend, Abelardo, she becomes a maid, a baby-sitter, and a prostitute. In Madrid, Dulce thinks of "each bed as a desk, a place of calculation, each body as a collection of unrelated parts.... Again and again I [Dulce] soldered myself into those deadening men, approximating lust" (285). A prostitute, Dulce's body becomes a sexual commodity in Miami, Havana, and Madrid. This complicates the question of the nation in the text. If women are alienated from their bodies to facilitate the mobility of the capital between nations, Dulce's story in that sense will echo Constancia's story, allegorizing transnational transactions to critique both communism and capitalism. Dislocation and involvement in oppressive and complex situations like Dulce's makes her story an allegory for the feminization of transnational labor that complicates national belonging.

Sections of the novel that take place in Cuba are marked by the years 1990–1991, which have been crucial in terms of the economic transformations in Cuba. One of Dulce's chapters (51–58) portrays the harsh years of the 1990s in Cuba. During these years, Cuba was denied the support of the World Bank and the International Monetary Fund as well as the opportunity to trade in American markets. In addition, the disintegration of the Soviet Union had decreased its economic support to Cuba. As part of making some market reforms, in the early 1990s, the Cuban government encouraged a sex industry by targeting tourists and promoting Cuba as AIDS-free. By the same token, despite the Cuban government's banning of public religious rituals, Santería rituals became more tolerated, specifically for foreigners "who want an authentic initiation" (56). Meanwhile, the Cuban government started to allow Cuban exiles to return to visit their family members. They bring "crammed suitcases" full of "photographs of ranch homes and Cadillacs, leather shoes... watches... extra-strength aspirin" (68). The abundance of this wide range of goods and everyday needs heightens the feeling of deprivation that Cuban citizens feel. Thus, the novel highlights economic and developmental stagnation in Cuba compared to the thriving economy that Cuban exiles enjoy in the United States. While the human cost of global capitalism is made visible in Cuba (Dulce is an example), it is totally invisible in Miami/New York where Constancia sells her Cuerpo de Cuba products. Despite the detailed description of the products, the factory, and the Santería rituals that Constancia resorts to for blessings, there is no reference to the workers, who works, working conditions, or even any reference to poverty or inequitable distribution of wealth in Miami/New York. As such, García draws a clear binary between first world capitalism and third world undeveloped nations. This in turn reemphasizes the one-way flow of monetary remittance from the sender (first world/United States) to the receiver (third world/Cuba). This binary makes the text resonate with Jameson's definition of a "national allegory" that invites interpreting the nation in terms of first world versus third world.

Unlike *West of the Jordan* in the second chapter where the monetary remittance (sent from the United States to Nawara in the West Bank) suggests an equal exchange with what Peggy Levitt calls "social" remittance (sent from Nawara to the United States) and unsettles the center–margin binary, *The Agüero Sisters* highlights that sender–receiver division. Although Santería rituals function as "social" remittance from Cuba and could balance the exchange of remittance between Cuba and the United States, these rituals are exploited in Miami and become part of an "affective economy," an emotional capital that circulates. Although readers do not have details, there is a glimpse of the Santero's "canary-yellow Buick" (192) and the aura of respect that accompanies his presence, which indicate that he is a public figure. This stands in contrast to the description of Santería rituals in Cuba where readers do not even have a description of a Santero figure. The metaphysical and the exotic render Santería more marketable not only among Cuban exiles but also within Cuba when it markets Santería beliefs for tourists, as I have mentioned earlier. Thus, the social remittance acquires a monetary value that circulates back to Cuba through tourists and foreign currency.

The Agüero Sisters has an element of magical realism, produced by the incorporation of Santería traditions and magical incidents such as Reina's survival of a fire accident and the changes that took place and changed Constancia's face to look exactly like her mother. Throughout the novel, these magical incidents revive the past in the characters' minds and emotionally connect them to Cuba. For instance, despite the tension between Reina and Constancia, both deeply believe in miracles and the metaphysical powers of Santería beliefs. The source of the tension between the two sisters is that each has a different version of how their mother died. Reina believes that it is an accident, and Constancia knows that it is Ignacio's revenge for Blanca's infidelity. In addition, as I have mentioned earlier, Constancia feels jealous that Reina has more access to their father's works and Reina feels jealous because she believes that Constancia knows what happened to their mother. These conflicts over the past reveal a desire for claiming ownership of the family's history in Cuba and by extension Cuba's cultural legacy (since the father's stories go back to Cuba in the nineteenth century). In a clear correlation between the magical and reclaiming the past, the Santero asks Constancia—as part of the blessings of the factory in Miami—to swim to Cuba and get hold of what is left of her father's legacy. This capture of the past is a clear signal in the text that their past/legacy and connection to the island brings Cubans together. Though this connection is essential, it is painful and hence repressed. In a confrontational scene between Reina and Constancia, when Reina wants to talk about their mother, Constancia covers her ears and "hum[s] the national anthem" (158). Constancia avoids feeling the burden of the past and the death of her mother, by covering her ears. Nevertheless, humming the national anthem stands for her deep involvement with that history of their mother.

García's use of magical realism as a narrative tool in *The Agüero Sisters* can be situated within the postcolonial approach to magical realism,[3] which defines the genre to be inherently historical, allegorizing and reflecting a historical crisis, precisely depicting colonial encounters and postindependence social dilemmas. García's text can be positioned within the colonial and postcolonial crisis of Cuba, without dismantling the third-world—first-world binary between Cuba and the United States that frames the novel.

The Night Counter

The Night Counter is the only full-length first novel by Alia Yunis, published in 2009. In 2011, Yunis published a short story titled "Girls on Ice" in the online literary magazine *Guernica*. Though *The Night Counter* has not been discussed in scholarly reviews and studies, it has been critically acclaimed by the press as "captivating," "comedic," and "rich in characterization." Kathryn Kysar writes in the *Minneapolis Star Tribune*: "'*The Night Counter*,' Alia Yunis' first novel, mixes equal parts of magical realism, social commentary, family drama and lighthearted humor to create a delicious and intriguing indulgence worth savoring" (August 2009).

In the novel's magical framework, Scheherazade reveals her presence to Fatima Abdullah for 992 nights. Fatima knows that once she completes 1000 nights of telling stories to Scheherazade she will die the night after, so she has to prepare for her death in ten days. In a role reversal, Scheherazade does not tell stories to Fatima but rather listens to Fatima's stories about her two marriages, her children, and her grandchildren. Throughout the novel, Scheherazade travels, following almost all members of Fatima's big family in the United States, Lebanon, or Gaza. The title of the novel on the cover page looks like Hindi letters rather than English or even Arabic. This Pan-Asian aura of the narrative is intensified by the images of mosaic patterns on the title page. In addition, the copyright page acknowledges citing parts of *The Arabian Nights*. Right after introducing the family tree of Fatima, some verses from *The Arabian Nights* are cited. These verses in *The Arabian Nights* are cited from Mu'allaqat (suspended Odes) Labid Ibn Rabia.[4] Written in the poetic tradition of the time, the Odes start with description of deserted dwellings and what is left of them. In light of Benjamin's insights about the form of allegory mediated through melancholia (where the subject feels the emptiness not only of nature but also its own ego, sees the face of death in nature rather than its beauty), one can infer that Jahilya poets, like Ibn Rabia, may be allegorists. In addition, not only are Ibn Rabia's poems allegorical but *The Arabian Nights* itself is an allegory in which stories of animals and human virtues and sins are all narrated to stand for something else beyond their surface meaning.[5] Benjamin's as well as Jameson's explanations of allegory speak to this text. Yunis, by conjuring up *The Arabian Nights* at the opening of her book, specifically these verses, is following a literary tradition of starting by lamenting emptied homes and their ruins. Yet, for an American audience this can situate the novel within the realm of exotic adventures and storytelling associated with the circulation of *The Arabian Nights* in Europe and America from the eighteenth century.[6] The cited verses are a plea to God to bring back those who have long left their dwellings. Hence, there is an implicit sense of loss, dispersal, and yearning for reconnecting in the opening of the novel.

The first chapter of the novel opens with Fatima going home after attending the funeral of Selma Haddad, an acquaintance. Since Fatima has moved to Los Angeles to stay with her grandson, Amir, and spend her last ten days with him, she decides to attend funerals of Arab acquaintances she knows in Los Angeles. This will help Fatima arrange her own funeral, leaving all the instructions for Amir. Fatima not only wants Amir to take care of all the details of her funeral, but also wants to find him a wife so that she can give him the keys to her house in Deir Zeitoon, in Lebanon. Fatima's persistent desires to find Amir a wife and to save the key of her house in Lebanon are odd, because on the one hand, Amir tells her that he is gay and he is not getting married, and on the other hand, Fatima has not been to Lebanon, or her alleged home, in decades. Thus, the Deir Zeitoon house is a deserted home, empty of its residents. Since Yunis does not give details of Fatima's house in Lebanon and does not depict any characters nostalgic to return to that house, it is a logical inference, then, that the house no longer exists after decades of living abroad. Such a conclusion is reached by almost all of Fatima's family as well as by Scheherazade; only Fatima believes that her house is still standing in Deir Zeitoon. At the end of the novel, her granddaughter, Dina, goes to Lebanon and starts looking for Fatima's house, but she finds out that the house no longer exists and the whole village has drastically changed. Dina calls Fatima and almost all her family members witness that phone call. When Fatima is asked by her daughter, Nadia, what her niece, Dina, is looking for, Decimal—Fatima's great granddaughter—answers on Fatima's behalf that Dina is "looking for the house with the terra-cotta roof, wood-burning stove, the cedar closets, and the four marble steps that lead to the top. . . . And the garden with lavender and jasmine and the fig trees all with figs" (353). Though she has never seen Deir Zeitoon's house, Decimal gives a full account of its beauty: lavender, jasmine, fig trees. Nevertheless, the house is gone. In addition, all the landmark houses and shops as well as the families who used to live in Deir Zeitoon are no longer there. For example, "The black-smith closed his shop *fifty years ago*, when all his sons went to America [.] . . . They own a hot dog shop in Cleveland" (354 emphasis added). The specific time reference in this quote alludes to—in Jameson's terms—the "embattled situation" in Lebanon that led to the migration of many Lebanese, namely, the colonial wars between the British and the French on the one hand and the Ottoman Empire on the other hand. In the early twentieth century, the three colonial powers—Ottoman, British, and French—divided the Levant region. Initially under Ottoman rule, the Levant region (Syria, Lebanon, Jordan, and the Palestinian territories) fell to the British and the French. Thus, Syria and Lebanon became French mandates, while the rest of the Levant became a British mandate. The French and the British were at war with the Turks. Fatima's father was killed in 1919 because he refused to join the army when the Turkish army used to put "Arabs" on the frontline in wars. Marwan, Fatima's first husband, was born when "Lebanon was *starving* because *Britain* and *France* had blockaded [Lebanese] harbors to defeat the *Turks*" (34 emphasis added). A puppet in the hands of colonial powers, Lebanon—as a nation—did not have any sort of sovereignty. Marwan was born at the time of the blockade, and Fatima was orphaned because of the Turks. Marwan and Fatima got married and left for the United States just before World War II. There is, hence, a correlation between Fatima's story of escaping death in Lebanon and the story of Lebanon, struggling for independence as a nascent nation.

However, Fatima's story is too complex to be simply read as a story allegorizing Lebanon, because it is also intertwined with stories of all her family members, who either live in the United States or live between the United States and Lebanon/Gaza. Fatima is the narrator of these stories and the central character. These stories are not arranged as chapters but as nights. For example, "The 998th Night" includes stories about Dina, Scheherazade, Amir, Fatima, and Decimal. The nights are chronological but the stories included in each night are not. Yet these stories are closely connected. Stories within stories is the narrative structure of *The Night Counter*, and it echoes the structure of *The Arabian Nights*: every night Scheherazade tells a story that she deliberately does not finish and resumes the next night. *The Arabian Nights* itself is an allegory in which stories of animals and human virtues and sins are all narrated to stand for something else beyond their surface meaning, as I mentioned earlier. By emulating the narrative form of *The Arabian Nights*, Yunis's story invites interpretation as an allegory, in which multiple individuals' stories reveal intricate political and economic relationships in the United States, and problematize the full integration of Arab Americans as American citizens.

I, therefore, argue that the non-national subject is a space of allegory. The two contradictory and contrasting stories of Randa (Fatima's daughter) and Amir (Fatima's grandson) bear this out. Despite their long presence, Arab Americans are still invisible, politically. Randa and her husband have helped the FBI to interrogate Arab Americans in Houston. Randa wants to show her neighbors that she and her husband are not terrorists. However, Randa realizes that "by offering to help them [FBI] I would let the neighbors find out that we were *Arabs*, and that just was worse than offering to be *patriotic*" (350 emphasis added). Amir is an amateur actor who, despite the fact that he looks for different roles to play, is always asked to wear a long beard to play the role of a terrorist or an Arab. Because of his costume beard he is under surveillance, following the attacks of 9/11. Fatima is under surveillance as well because she wears a headscarf. Had the FBI agent, Sheri Hazad, seen beyond the beard and the headscarf, she would have known that the beard is false and Fatima is too old to be a national threat. In this sense, Amir and Fatima fit into what Leti Volpp calls in "The Citizen and the Terrorist" (2002) the "new identity category that groups together persons who appear 'Middle Eastern, Arab, or Muslim.'" As she elaborates, "members of this group are *identified* as terrorists, and are *dis-identified* as citizens" (1576 emphasis added). The FBI agent "identifies" Amir and Fatima in light of their appearance and "dis-identifies" their rights as American citizens by violating their privacy. Furthermore, had the agent seen beyond religious symbols, she would have known that Amir is gay and that homosexuality is abhorred among orthodox Muslims. Randa, Fatima, and Amir reveal that Arab and/or Muslim citizens are perceived as the foreigners or enemies within, threatening America's security. The Patriot Act renders Arab Americans and/or Muslims non-nationals. By virtue of its name, The Patriot Act defines patriotism in terms of enacting measures such as surveillance, detention, and in some cases deportation of individuals who are suspected to be a threat to the United States. Thus, the Act by its very definition is an exclusive act that turns individuals against each other to perform the role of "protectors of U.S. politics," as Donald Pease points out (2009, 31). The Patriot Act aims at bringing together Americans, yet the only means to stand together as Americans is to exclude other Americans who look, sound, or seem, Arab/Muslim. Randa and Amir, though born in the United States, are outsiders in the dominant American national discourse.

Like the Patriot Act, the war in Iraq generates paradoxical meanings of patriotism. The war in Iraq, allegedly a war on terror, implies the ultimate sense of patriotism: sacrificing one's life for the sake of one's nation. Rock, Miriam's son, decides to go to Iraq and before him, Joseph, his father, served in Vietnam. Miriam, however, is reluctant to let her son go to Iraq, because she does not want to face the devastating consequences of a war. Miriam's mixed feelings are shared by the congregation of her church where they start what might seem paradoxical charitable acts, yet humanitarian in a universal sense. Miriam has started a clothes' drive for Iraqi orphans, and simultaneously raises funds to help with college fees for US soldiers' children (256). Amir's and Rock's stories complicate the meaning of collectivity within the Patriot Act and the war on terror.

The meaning of a *collectivity* as a *community* is likewise problematic in Yunis's text. *The Night Counter* offers a series of collective identities rather than *a* collective identity for Arab Americans in the United States. Katherine Sugg observes that many US ethnic and minority literary texts function "allegorically with a protagonist-narrator who embodies . . . the history and collective subject of the community" and the "nation" (30). Arab American collective identities in *The Night Counter* differ from Sugg's position. While it is true that Amir's dilemma may allegorically stand for the political status of Arab Americans within the United States, the story of Sheri Hazad (the FBI agent) does not. On the contrary, Sheri Hazad's story shows full integration of an Arab American within the United States to the extent of bearing the responsibility of maintaining US security. It is only when we read Amir's experience together with Sheri's experience that we can see that neither character stands for all Arab Americans. In the same way, neither Randa's cooperation with the FBI nor Rock's decision to join the American troops in Iraq stand for all Arab Americans. I, therefore, pluralize the collective subjectivities in Sugg's proposition. Amir, Sheri Hazade, Randa, and Rock complicate the possibility of a singular collective subjectivity for Arab Americans. Rock, Randa, and Sheri stand for what the American administration would count as "good Muslims." Mahmood Mamdani in *Good Muslim, Bad Muslim: America, the Cold War, and the Roots of Terror* (2004) analyzes the distinction that President Bush made between good Muslims and bad Muslims after 9/11: "'good Muslims' were anxious to clear their names and consciences of this horrible crime and would undoubtedly support 'us' in a war against 'them.'"

As Mamdani concludes, "[j]udgments of 'good' and 'bad' refer to Muslim political identities, not to cultural or religious ones" (15). Miriam's character, however, depoliticizes Islam and disrupts the good Muslim–bad Muslim binary. On the one hand, despite raising funds to help with college fees for US soldiers' children, she has also started a clothes' drive for Iraqi orphans (256). On the other hand, Miriam is a Muslim married to a Christian man and is involved in the congregation of his church. That is to say, Miriam's double allegiance crosses political boundaries and religious differences for the sake of humanitarian goals.

Not only does the FBI agent see Amir as a potential terrorist, but Hollywood also perpetuates the stereotypical representations of Arabs on its screens. In the phone call between Amir and his agent talking about a future audition, the agent calls Amir "Osama" (26). In addition, though asked to play the role of a cab driver in New York, not a terrorist, he is asked to wear his long beard. In Hollywood, Amir is not only marked by his beard but also has to conceal his sexuality. If producers in Hollywood know he is gay, "they'll never let you [Amir] audition for the terrorist parts" (26). In this sense, Amir is reduced to his beard. On the one hand, Amir's beard is a national threat, so he is targeted to sustain the nation's security. On the other hand, in Hollywood, his beard fulfills the demand for actors with Middle Eastern features but he has to conceal his sexuality to be convincing as a terrorist.

The fetishization of the Muslim beard and headscarf by the FBI agents and Hollywood producers is a reflection of perceiving Amir and Fatima as "objects of fear" (65), as Sara Ahmed would say. Analyzing the "affective politics of fear" in general and within the context of the war on terror in specific, Ahmed holds that the security measures that were taken in the aftermath of the September attacks allowed unjustified detentions, deportations, and wars on other nations, a narrative that Ahmed calls a "could-be terrorist" narrative. This narrative, as she elaborates, "relies on the structural possibility that the terrorist 'could be' anyone and anywhere" (79)—evoked by the "could-be [terrorists]" narrative that has circulated after 9/11 attacks about Arab Americans and

Muslim-Americans. The "could-be" narrative is a paradoxical narrative in which the potential terrorist could be anywhere but not in one specific place. Ahmed argues that the "affective politics" of fear aim at preserving "through announcing a threat to life itself" (64). This fear is paradoxical because as Ahmed argues, it "involves relationships of proximity" but it "re-establishes distance between bodies whose difference is read off the surface" and "involves the repetition of stereotypes" (63). Despite her proximity to Amir and Fatima, the FBI agent does not see beyond the stereotypical correlation between Islam and terrorism. In that sense, Amir and Fatima are, in terms of Brian Massumi, "shadowy figure[s]" (qtd. in Ahmed 79), associated with terrorism. These paradoxical dynamics of fetishizing the terrorist body are also accompanied by paradoxical dynamics of racializing the terrorist figure. Jasbir Puar in *Terrorist Assemblages: Homonationalism in Queer Times* (2007) holds that not only the terrorist figure is racialized as either Middle Eastern or Arab but also the terrorist body "must appear improperly racialized (outside the norms of multiculturalism) and perversely sexualized in order to materialize as a terrorist in the first place" (38). Thus, in light of Puar's argument about the interrelationship between US national citizenship and heteronormativity that implicitly excludes lesbian, gay, and queer bodies, Amir's homosexuality reemphasizes his exclusion. The stories of Fatima, Amir, Randa, Rock, Sheri Hazad, and Miriam are allegorical stories of the non-national subject formed at the nexus of *disidentification* and the *could-be terrorist narrative.*

The Night Counter does not necessarily foreground double allegiance, like *The Agüero Sisters*, but rather dismantles the conflation of Arabs and Muslims. For example, Zade, Fatima's grandson, owns an international dating business called "Scheherazade's Diwan Café." He and his partner, Giselle, started this business as a hookah place and then they expanded their business to add an international dating business, Aladdin and Jasmine, Inc. The dating business grows internationally to include Muslims worldwide: Afghani, South Asian, Turkish, Bosnian, Lebanese and Syrian Armenians, and Black Muslims (60). This expansion demonstrates that not all Arabs are Muslims and not all Muslims are Arabs. The questionnaire about the perfect partner is comprehensive in a sense that makes it possible to include all sorts of Arabs/ Muslims, yet, in the attempt to be inclusive and comprehensive, the questionnaire shows that being Arab or Muslim is not solely regional or racial, but rather hinges on cultural traditions, political views, or simply personal preferences. For example, a Qatari man, though already married, is looking for a second wife who ideally should be "a nice Arab-American bilingual highly educated virgin not opposed to wearing the *abaya* and conversant in French cuisine" (46). Though Zade wants to dismiss the profile of the Qatari man, he does not want to reject his business. Hence, Zade simply adds another question to the questionnaire: "How do you feel about being a second wife? a) abhorrent b) mildly abhorrent c) acceptable if there is no other option and I still want to get married" (48). A girl who is bilingual, highly educated, virgin, conversant in French cuisine need not be necessarily Arab or Muslim. Likewise, accepting polygamy is not intrinsically Muslim, because many Muslim women are against polygamy. Along the same lines the questionnaire asks about voting, the war in Iraq, helping in-laws to migrate to the United States, eating pork, and/ or wearing hijab/abaya/niqab or using sunscreen instead of wearing hijab. Comparing hijab to using sunscreen takes away the religious meaning and symbolism of hijab and renders it merely an extra layer that can be replaced by sunscreen. The question makes fun of Muslim men and women who believe in hijab, yet use sunscreen as a substitute. In the same way, the question about pork and alcohol makes fun of Muslims who eat pork in secret or do not mind eating one specific form of pork. Though both questions address Muslims sticking to Muslim teachings without questioning them, these questions are also addressing mainstream discourse about Muslims that represent them as a unified, fixed, homogeneous collectivity.

The questionnaire for dating Arab men and women is not only a critique of the homogenization of Arab men and women, but it also reveals that there is no consensus among Arabs themselves on what is authentically Arab/Muslim; it varies, changes, and transforms according to cultural context. One of the questions is what type of Arab the applicant is interested in dating: "Mediterranean—Egyptian—Persian Gulf—North African—East African" (48). The obvious fact that there are no clear-cut bases for these distinctions among Arabs means that it is not reasonable to put all the Arabs in one box as one homogeneous group. Decimal—Fatima's great granddaughter—is an example; she is a mix of an Arab-American mother and a Brazilian father. Her Brazilian father is from Syria; he migrated to Brazil, married a Brazilian wife, and hence became Brazilian. Decimal describes herself as "one-fourth Muslim...50 percent Christian and 25 percent Taoist" (303–304). Decimal does not identify as solely Arab or Muslim (even though she is perceived as only Arab). Mocking racial classification in the United States, Decimal says: "if Paolo's [her boyfriend] Syrian and I'm one quarter Lebanese, my baby's going to be three-eighth more Arab than me and just three eight less Arab than you [Amir]" (291). The name Decimal itself is metaphoric. A decimal number is two digits, implicitly, then, the name Decimal suggests openness to numerous combinations. Yunis's portrayal of a wide range of Arab Americans disrupts the monolithicization of Arab American as an ethnic group, which can be read as a counter-argument to the demonization of Islam in the aftermath of the attacks in September 2001.

Intertwined with the political axis in the novel is the economic axis that renders *The Night Counter* a transnational allegory. Like *The Agüero Sisters*, *The Night Counter* portrays some aspects of capitalism in the United States. Yet *The Night Counter* depicts economic inequalities within the United States, rather than within postindependence nations, like Cuba that suffer the brutality of capitalism, which devours their economic systems.

Fatima's arrival scene in New York in the 1930s reveals an economic imbalance between those who own means of production and those who work for the owners. Upon her arrival in New York, Fatima was amazed by its extremely tall buildings, thinking that "something so far up could not stand by itself for so long, and

I [Fatima] kept covering my head, anticipating one to fall" (81). Fatima's anticipation of danger is tied to her witnessing the hardships of living during the depression years. She says:

> When I finally opened my eyes and looked *ahead instead of up*, people stood in long lines for bread that was hard and thick, and shop clerks chased kids in rags for stealing apples. (81 emphasis added)

The contrast between what Fatima sees when she looks up (the illusionary sense of luxury that New York skyscrapers evoke) and what she sees when she looks ahead (the dire living conditions that most people struggle with in order to survive) calls attention to an economic imbalance between laborers and those who own means of production. This economic imbalance has expanded outside the United States, especially after the discovery of oil in the 1970s in the Arab Gulf states and the circulation of "petrodollars." Ibrahim Oweiss in "Petrodollar Surplus: Trends and Economic Impact" (1984) coined the terms "petrodollar" and "petrodollar recycling" in his analysis of the economic investments between the United States, the UK, and oil-exporting Arab nations. He explains that oil is priced in US dollars, so "petrodollars" are US dollars earned through international oil trade (177). Recycling petrodollar flows means Arab Gulf states that produce oil and export oil, own a surplus of dollars, which they reinvest back in the United States (179). Consequently, "petrodollar" investing nations become richer, which leads to a global economic disequilibrium between oil-investing nations and nations without oil resources.

Scheherazade's Diwan Café, Aladdin and Jasmine, Inc., and Arab tourism in Las Vegas are the sites in the novel for recycling petrodollars. The contradiction between looking up and looking ahead, which Fatima has seen in New York is echoed in Vegas. While Saudis spend their money on alcohol, gambling, belly dancers, and prostitutes, other Arabs who are not from Arab Gulf states work as cab drivers. The novel includes the stories of Bassam, Hossam, Nassim, and Wissam who live at the margin of Vegas's luxurious nightlife. Bassam could not be an engineer as his parents wished. Hossam's family left Lebanon because of wars and he could not finish his law degree in the University of Jordan. Nassim was an engineer in Tunisia but could not find a suitable job in the United States. Wissam was a doctor in Algeria who left for the United States to pursue the American Dream. Rather than sharing their success stories, they share their "geopolitical sob stories" and how they were professionals in their home countries (310, 306). Furthermore, after 9/11 and, in order to fit in, all the Samirs, Wissams, Osamas, Hossams, and Nassims shortened their names to Sam. Thus, Bassam is Sam; Hossam is Sam #2; Nassim is Sam #6; Wissam is Sam #17. In allegorical terms, the story (ies) of Sam (s) replace that of Uncle Sam—America's national symbol. The new Sam does not stand for American ideals of freedom, equality, and justice, but rather for disappointment and frustration. Unlike Constancia's business that conceals the human cost of her business, stories of cabbies in Vegas point out the economic imbalance between the Saudis who spend their money in Vegas and the cabbies who work for the Saudis *in* the United States.

Zade and Giselle's Scheherazade's Diwan Café and Aladdin and Jasmine Inc. started with borrowed money from Giselle's brother who works in Saudi Arabia and from Zade's Qatari brother-in-law. After 9/11, both businesses boomed as more Arab men and women started looking for Arab partners. Nevertheless, that was not the sole reason for that business to boom. Another reason, as Zade's partner—Giselle—puts it, is that a hookah/ dating place, named Scheherazade in D.C. is "catching a trend," meeting a need (52). It is a place where "love cashes the checks" (55). Zade's business expands and grows into an international business that is not only for Arab men and women in the United States, as I have clarified earlier. In the light of Giselle's reasoning, Zade's business relies on the circulation of the figure of Scheherazade.

The expansion of Zade's dating business beyond the geographical boundaries of the United States is intertwined with orientalizing this business by adding the name "Scheherazade." The interest in "creating and maintaining a sense of radical difference between West and East" reemerges especially after the September attacks, as Saree Makdisi and Felicity Nussbaum point out in *The Arabian Nights in Historical Context: Between East and West* (6). On one level, orientalizing a business in the era of global capitalism echoes the stereotypical implications of the orient as exotic, erotic, as well as traditional and backward. The novel partially resonates with these stereotypes. On the walls of the café, Scheherazade is portrayed as a half-naked belly dancer. In Vegas, belly dancers are named after Scheherazade: "the scintillating Scheherazade" (313). On another level, I would like to interpret Scheherazade as a non-national figure. Quoting André Gunder Frank, Susan Friedman, in "Unthinking Manifest Destiny: Muslim Modernities on Three Continents" (2007), underlines that the rise of the West was enabled by "borrowing from other cultures" and the "influx of gold and silver," which allowed Europe to "compete in the global market" (70). Friedman concludes that the modern West is "derivative," "depending for its rise...on global factors and the long-standing transcontinental exchange of goods, technology, peoples, and cultures" (71). Scheherazade is a mobile character in the novel, traveling between the United States, Lebanon, and Gaza. In a similar sense, the café that is named after her is global, symbolizing her mobility beyond geographical boundaries and unfolding several political and economic exchanges in the novel. No fixed origins is precisely what *The Arabian Nights* as a cluster of stories evokes; it is Arab, Indian, Persian. Hence, Scheherazade is not confined to specific national territories, a non-national figure in this particular sense, and can be read as a *signifier* with multiple meanings.

Though the novel acknowledges Scheherazade, the erudite woman, by citing some verses from *The Arabian Nights*, the rest of the novel shows Scheherazade traveling between the United States and the Arab region, not telling stories but listening to Fatima's stories, eye-witnessing different stories and incidents in the lives of Fatima's children, grandchildren, and grand grandchildren. Scheherazade's character can be read allegorically for the conflict between the old and the new, East and West. Scheherazade's reappearance to Fatima renders the novel an allegorical interpretation, representing the clash between the East, where Scheherazade belongs (read: precapitalist, old, and magical) and the West, where she moves to (read: capitalist, new, and real). The outcome of this clash is the *death* of Scheherazade, symbolized in the novel by Fatima's impending death. Scheherazade is the one who is listening to Fatima's stories, not the one telling them. The *Arabian Nights'* witty, sharp, charming storyteller is visible only to Fatima, the other characters only see images of Scheherazade as a half-naked belly dancer on the walls of Zade's café. Zade manipulates the old to catch up with the benefits of the new, hence fragmenting Schehera/Zade and transforming her intellectual legacy—which appears on the first page of the novel before the story starts—into a sex symbol. What is more significant about *The Night Counter* is that it generates a complex "signifying process" that blurs boundaries between nations, and hence opens the novel to multiple meanings and interpretations.

To conclude, reading together *The Night Counter* and *The Agüero Sisters* I interpret both texts as national and transnational allegories and suggest the possibility of interpreting the non-national subject allegorically. Both texts, written in the United States, demonstrate three historical crises which the United States has engaged with—the rise and fall of communism, the conflicts caused by capitalism, and the 9/11 attacks—which reveal the overlapping histories of the U.S. nation-state and the Arab/Muslim countries, and Cuba. This overlap of national destinies is present in Yunis's and García's texts, but produces two different effects. A point of convergence between *The Night Counter* and *The Agüero Sisters* is the circulation of capital (Zade's dating business and Constancia's cosmetic products) that blurs geographical boundaries: New York is the site of devastating consequences of capitalism (*The Night Counter*); and Havana and Madrid are the exploitative territories of women's labor (*The Agüero Sisters*). Nonetheless, while *The Night Counter* dismantles national boundaries, *The Agüero Sisters* recapitulates the economic boundaries (and possibly advantages) between the United States and third world nations that struggle within neoliberal economic systems.

Afterword

In this book, I focus on the implications of the non-national in narratives by Arab-American, Chicana, South Asian-American, and Cuban-American women writers. As I mentioned earlier in the introduction for this study, the non-national subject reconfigures implications of home country and host country in the migrant stories I have discussed. The non-national, however, should not be confused or conflated with the transnational; it is rather a specific moment within the transnational that problematizes both the national and the transnational, and complicates meanings of national consciousness, national time, national space, and national belonging as the earlier chapters suggest. That is to say, the non-national subject contests and questions unproblematic assimilation and integration into the United States (as the examples of *The Language of Baklava* and *An American Brat*); and resists singular belongings and binaries between first world and third world; home and abroad (as the examples of *The Night Counter* and *The Agüero Sisters* clarify). As I have mentioned earlier, I expand the meaning and use of the non-national in American Studies. The analysis of the texts in this study shows that the non-national opens up narratives to multiple interpretations to rethink Americanness in light of intricate US domestic and foreign policies.

Texts in my study share themes of migration and/or problematic belongings. However, themes of food and uneven economies also recur, which open up the possibility of comparing these texts from multiple angles. For example, *The Language of Baklava* and *The Namesake* can be read together as they share the theme of food as a cultural symbol that problematizes national affiliations. Similarly, *An American Brat* and *The Agüero Sisters* can be analyzed in light of the consequences of capitalism in Third World countries such as Pakistan and Cuba. That is to say, though the writers in this book have different histories in the United States, a contrapuntal reading brings them together, yet maintains the uniqueness of each text rather than seeking their similarities. A contrapuntal reading gives a way of understanding the moving between different spaces—the United States, Jordan, Pakistan, the West Bank, Mexico, Egypt, India, Lebanon, and Cuba—to contest monolithic understandings of Americanness and assimilation. Reading contrapuntally offers a broader perspective to analyze the notion of the nation, national consciousness, national time, national space, and national belonging not only in the United States but also in the native countries of the writers in my study. Furthermore, this particular way of reading brings out the full diversity of Arab-American women's writing.

The goal of this book is not merely drawing attention to Arab-American women's writing and demanding their inclusion in more literary anthologies. My aim is twofold. First, I expand, complicate, and open new questions about the meanings and use of the term "non-national" within new Americanists' studies. Second, in my study I have emblematically used narratives by Chicana, South Asian-American, and Cuban-American women writers to underline what Arab-American women writers share and do not share with these writers. By comparing Arab-American texts to the Pakistani-American, Chicana, Indian-American, and Cuban-American texts, I underline that Arab-American narratives can potentially expand the meanings of the non-national without essentializations of or generalizations about both terms, Arab American and non-national.

Notes

Introduction

1. Frederick Crew is one of the first to describe—and resist—the "new Americanists" who brought a change to the field of American Studies in his article "Whose American Renaissance" in (1988) (qtd. in Pease *National Identities* 1).
2. Rob Wilson uses this description, in "Techno-euphoria and the discourse of the American sublime," to refer to the expansionist decades of manifest destiny (1835–1855) (206).
3. I borrow here Justice Edward Douglass White's definition of "foreign in a domestic sense" when he explained how Puerto Rico, Guam, and the Philippines are not incorporated into the United States after their acquisition from Spain. They belong to, but are not part of the United States (qtd. in Christina Duffy Burnett and Burke Marshall 1).
4. Arabic-speaking countries include North African countries (Egypt, Morocco, Tunisia, Algeria, and Libya); West Asian countries (Iraq, Kuwait, Saudi Arabia, Bahrain, Jordan, Lebanon, Qatar, Syria, Palestinian-occupied territories, United Arab Emirates, and Yemen); Mauritania (in West Africa); and Somalia and Djibouti (in Northeast Africa).
5. For the history of Greater Syria, I rely on *Greater Syria* by Daniel Pipes (1990). Pipes explains that before World War I, Greater Syria "refers to a region stretching from the borders of Turkey to those of Egypt, from the edge of Iraq to the Mediterranean Sea" (2). After Word War I, Greater Syria was divided into Syria, Lebanon, Jordan, and Palestine (3).
6. The website bintjbeil.com, setup by Arab American community in Detroit, Michigan, has a section titled, "100 Questions and Answers about Arab Americans." It explains that a hyphen is used when "Arab-American" is used as adjective, but without a hyphen when referring to someone who is Arab American.
7. I use the term "racialization" in the same sense suggested by Karim Murji and John Solomos in *Racialization: Studies in Theory and Practice* (2005). Namely, racialization "refers both to

ideological practices through which race is given significance, and cultural or political processes or situations where race is invoked as an explanation or a means of understanding," as Murji and Solomos put it (11).

8. Robert Miles coined the term, "racialization of migrant labor" in *Racism and Migrant Labour* (1982) to analyze racist discriminations in England against migrant workers from the Caribbean, India, and Pakistan, especially in the 1960s and 1970s, confining them to unskilled manual jobs (168–171).
9. According to the US census of population, people from the Middle East and North Africa are classed "white." As stated by the US Census Bureau, "white" is a "person having origins in any of the original peoples of Europe, the Middle East, or North Africa" (quickfacts.census.gov/qfd/meta/long_68178). Mathew Frye Jacobson gives a historical account of the constructedness of Caucasian "whites" and non-Caucasian "whites" in *Whiteness of a Different Color: European Immigrants and the Alchemy of Race.* He gives different examples of immigrants from some parts of Europe, Jews, and Syrians who were naturalized by court in early twentieth century as "white persons" (237–243).
10. Examples for these anthologies are: *Talking Visions: Multicultural Feminism in a Transnational Age* (Ella Shohat, 2001); *Colonize This!: Young Women of Color on Today's Feminism* (Daisy Hernandez et al., 2002); *This Bridge We Call Home: Radical Visions for Transformation* (Gloria Anzaldúa and Ana Louise Keating, 2002); *The Color of Violence: The Incite!* (Incite! Women of Color against Violence, 2006).
11. A relevant example in this context is Laura Bush's weekly presidential address in November 2001 in which she underscores: "Because of our recent military gains in much of Afghanistan, women are no longer imprisoned in their homes" (qtd in Ann Brodsky 116).
12. This conflation has itself marginalized Arab-Jewish feminists such as Ella Habiba Shohat (an Iraqi-Israeli) and Kyla Wazana Tompkins (a Jewish Moroccan) because they are Arabs and non-Ashkenazi Jews.

1 The Non-national Subject in *The Language of Baklava* and *An American Brat*

1. Theodore Roosevelt held that ethnic loyalties, if they are necessary for some groups, should be only subordinate to the higher unity of the nation (qtd in Yehoshua Arieli 187).

2. I will clarify later in my discussion the complexity of the racial category of people from the Middle East as white.
3. Tracing the genealogy of the term, Raymond Williams holds in *Keywords: A Vocabulary of Culture and Society* (1983) that "ideology" has been used to mean "the set of ideas which arise from a given set of material interest, or more broadly, from a definite class or group" and is used paradoxically to mean false consciousness (because it serves the interests of a specific class or group) (155–156).
4. I am referring here to Louis Althusser's definition of interpellation, in his essay "Ideology and Ideological State Apparatuses" in *Lenin and Philosophy* (1971), as a process of hailing/addressing individuals as subjects (170).
5. Diana's experience is somewhat different from Feroza's. Diana's mother is American and does not appear to force her daughters to be more "American" than Jordanian. In contrast, Diana's father insists that his daughters are and should be Arabs/Jordanian.
6. In "Transmissions Interrupted: Reconfiguring Food, Memory, and Gender in the Cookbook-Memoirs of Middle Eastern Exiles" (2002), Bardenstein explains that there are subgenres and variations to the cook-book memoir genre: memoirs with recipes, culinary memoirs, nostalgic cookbooks (357). I consider Abu-Jaber's food memoir a subgenre of cookbook-memoir.
7. I am referring specifically to reviews by Devon Thomas, Pat Bangs, Gillian Engberg, and Joy Harris.
8. Based on a survey conducted in 2002 by the Arab American Institute in Washington DC, the religious affiliations of Arab Americans are as follows: 35 percent are Roman/Eastern Catholic, 18 percent are Eastern Orthodox, 10 percent are Protestant, 24 percent are Muslim, and 13 percent are of other religions or have no affiliations (www.aaiusa.org/arab-americans/22/demographics).
9. "The Brown DeMone" is a comic cartoon show that was popular in the 1960s and known by its vampire-like character.
10. Karen Leong, in *The China Mystique: Pearl S. Buck, Anna May Wong, Mayling Soong and the Transformation of American Orientalism*, holds that "Orientalism in the United States had its roots in the attitudes and values of European immigrants who arrived in North America during the sixteenth and seventeenth century.... [and] took a form specific to and supportive of the United States' emerging role as a world-wide moral and economic force" (2005, 7).

11. Hamid Naficy uses this phrase in *The Making of Exile Cultures: Iranian Television in Los Angeles* (1993) to describe the circulation of souvenirs, symbols, and icons from Iran among Iranian exiles—especially from parents to their children—to transmit to their children their Iranian heritage. The meanings that "cultural mnemonics" produce involve "establishing both cultural and ethnic differentiation (from the host city) and cultural and ethnic continuity (with an idealized past and the homeland)" (151–152).
12. Shangri-la is a fictional idyllic place in James Hilton's novel *Lost Horizon* (1933).
13. In "Disjuncture and Difference in the Global Cultural Economy" (1996), Appadurai defines "mediascapes" to be "the distribution of the electronic capabilities to produce and disseminate information (newspaper, magazines, television stations, and film-production studios)." He also defines "ideoscapes" to be "concatenations of images, but they are often directly political…[consist] of a chain of ideas, terms, and images, including *freedom, welfare, rights, sovereignty, representation*" (35–36).
14. I am referring to reviews specifically by Robert Morace, Winifred Sihon, Adam Penenberg, Eliza Bent, and Adele King.
15. Roots of Parsees are in Persia—what is now known as Iran. The Arabs' invasion of Persia in the seventh century coerced Parsees to renounce the Zoroastrian faith and convert to Islam. They fled to the Indian subcontinent and congregated in Bombay, Lahore, and Karachi; after the partition of the Indian subcontinent in 1947, some have stayed in India and Pakistan, and some are diasporic.
16. Jürgen Habermas's study, *The Structural Transformation of the Public Sphere* (1962), is a genealogy of the bourgeois public sphere, focusing on its social and economic bases. Habermas's accounts for the public sphere as a domain of "critical judgment" of state and government (24).
17. For this background information, I rely on *Good Muslim, Bad Muslim: America the Cold War, and the Roots of Terror* by Mahmood Mamdani (2004). During the Cold War years with Russia, the United States supported Muslim Afghan insurgents who were fighting the spread of communism in Afghanistan. After the victory of the Saur revolution 1978 that brought communist parties to power, the Afghan fighters fled to Pakistan and were trained by the CIA. This was also accompanied by a shift in the US politics toward Pakistan. Despite Pakistan's violations of international humanitarian laws and the execution of Zulfikar Ali Bhutto, the United States offered huge financial and military aid

to Zia-Ul-Haq to host and support Islamist militants who joined the Afghan fighters from all over the world in their battle against communism.

18. I am borrowing from Japtok's study *Growing Up Ethnic: Nationalism and the Bildungsroman in African American and Jewish American Fiction.*

2 Reimagining the US National Time in *West of the Jordan* and *The Last Generation*

1. I rely here on Donald Pease's introduction to *The Futures of American Studies* (2002) in which he responds to and critiques Gene Wise's largely cited article "'Paradigm Dramas' in American Studies: A Cultural and Institutional History of the Movement."
2. In *Heritages of Our Times* (1991), Bloch examines the changes brought by capitalism to peasant life, which render peasant life contradictory, characterized by both traditional and contemporary and modern aspects. Bloch specifically defines "non-contemporaneity" as "unsurmountable remnants of older economic being and consciousness" (106).
3. I refer here to specific reviews by Parama Sarkar, Fay A. Chadwell, Dori DeSpain, and Elsa Gaztambide.
4. I am borrowing here from both Jeanette Winterson and Paul Ricoeur. Winterson uses "spiral narrative" to describe her first novel *Oranges Are not the Only Fruit* (1985). She explains that a spiral narrative technique is "fluid and allows infinite movement." It also reflects our mental and reading process: "every turning yields another turning" (xiii). Susana Onega describes the spiral narrative technique to be one that gives a sense of infinity (31). "Emplotment" or "configuration" is Ricoeur's term for the order of events in a novel. It is also what he refers to as "mimesis 2" in his schemata for fictional narrative in *Time and Narrative*, volume 1 (52–73).
5. Meyerhoff explains that these developments have made the world "technologically one," in the sense that what "happens now here, happens now everywhere; while we are at one place, we are potentially (with the negligible difference of a few hours) anywhere in the world" (109). Thus, these developments have "shortened time and expanded space" (111).
6. Peggy Levitt defines "social remittance" to be "the ideas, behaviors, identities, and social capital that flow from receiving-to sending-countries communities" (927). These transnational

exchanges sustain the family ties between relatives who leave and relatives who stay.

7. Page in her argument about Caribbean narratives as "cultural remittance" borrows the economic definition of remittance "as a one-way street where the [Caribbean] diaspora is implicated in positions at the center" (11).
8. Focusing specifically on Caribbean texts, Page explains that the "remittance text" is "the text that highlights, whether in celebration or critique, the inequalities of the exchange between Caribbean denizens of the metropolitan diaspora and Caribbean people who live in the region, traditionally separated as denizens of 'First and Third' world spaces respectively" (82).
9. I am drawing here from Anderson's definition of the nation as a community that is imagined. It is imagined because "in the minds of each [member of that community] lives the image of their communication." It is a community because it is "always conceived as a deep, horizontal comradeship" (7).
10. The West Bank was part of Syria under Ottoman rule (between 1517 and 1917), then became Palestinian by British mandate (1920). Later, it was annexed by Jordan (1948) and then Israel gained control over it after the 1967 war. Since 1993—after the Oslo Accord—parts of the West Bank are under Israeli control and other parts are under Palestinian control.
11. Mawal means a folkloric ballad that has different themes—female beauty, heroism, unfulfilled love stories, and separation. These ballads are popular because they are in common dialect and are transmitted from one generation to the next.
12. Marianne Hirsch in *The Mother/Daughter Plot: Narrative, Psychoanalysis, Feminism* (1989) coins her phrases "female family romance" and "feminist family romance" based on Nancy K. Miller's argument that the process of resistance and revisions in the works of women writers should be identified as "feminist." However, Hirsch limits the use of "feminist" to writings by women writers who specifically deploy ideas of the feminist movement in the 1970s and 1980s—especially psychoanalytic revisions of the family paradigm. Hirsch uses "female" for earlier works by women writers that include women's "self-consciousness and resistance" (8–15). For the purposes of my argument within the context of Halaby's novel, I will use "feminist" in the sense of resisting a dominant patriarchy not necessarily revisiting the psychoanalytic principles of a patriarchal family structure.

13. In the mid-1970s, Chicano writers like Alurista (pseudonym of the Chicano poet and activist Alberto Baltazar Urista Heredia) and Rudolfo Anaya depict in their writings a pre-Columbian Mexico, an Amerindian land (Aztlán) and its Aztec culture and mythology. Thus, Aztlán became a mythical cultural homeland for the Chicanos.
14. Coyoloxauhqui is the daughter of Coatlicue and sister of Huitzilopotchi—god of war—who, upon her birth, disremembered her and exiled her to darkness. Thus, he became the sun/son and she became the moon/daughter.
15. Moraga in this part of her book uses the precolonial name "Aztlán" to refer to the Bay Area.
16. "Queer Aztlán," however, runs the risk of essentializing Chicano/a national discourse. For example, Rafael Pérez-Torres in "Refiguring Aztlán" (2000) and *Movements in Chicano Poetry* (1995) addresses the problematic implications of reclaiming Aztlán to be the indigenous Chicano/a homeland. On the one hand, though Aztlán is perceived as a unifying symbol, its conceptualization "erases the vast differences that help for the richness and variety of the terms 'Chicana' and 'Chicano'" (115 "Refiguring Aztlán").
17. Edward Said in "Yeats and Decolonization" argues how imperialism is "an act of *geographical violence* through which virtually every space in the world is explored, charted, and finally brought under control." Therefore, for the native "the land is recoverable at first only through the *imagination*" to "reclaim, rename, and reinhabit the land" (225–256 emphasis added).
18. In the first poem in the book—"En Route para Los Angeles (On the Way to Lost Angeles)"—Moraga refers to New York as "Nueva York" and elaborates that "the name doesn't sound right/even spanglicized" (13).
19. Anders Stephanson identifies three historical moments in US foreign policies of "manifest destiny": seizing half the territory of Mexico in 1840s, adding the Philippines as an overseas colony after the Spanish-American war at the turn of the twentieth century, and fighting "forces of the communist evil" at the Cold War era (xiii, 119–121). The post–Cold War era marked the beginning of "deterritorialized capitalism," which decentralized capitalism so that no specific nation or region is the center of global capitalism. Euro-American "global capitalism" became the new economic system of controlling the world (129).

20. It is an interesting coincidence to see Moraga giving the example of the Intifada of Palestinians as an example for nations of people (169).

3 Moments of (Un)belonging: The Spatial Configuration of Home(land) in *The Time between Places: Stories that Weave in and out of Egypt and America* and *The Namesake*

1. Rob Wilson uses this description, in "Techno-euphoria and the discourse of the American sublime," to refer to the expansionist decades of manifest destiny (1835–1855) (206).
2. Kandiyoti elaborates that enclosure "encompasses racialized spatial segregation and immobilization *and* literary modalities that 'enclose'; that is, they center around discursively bordered, particularized loci, such as regionalism and urban writing" (5).
3. Brah defines "diaspora space" as "the point at which boundaries of inclusion and exclusion, of belonging and otherness, of 'us' and 'them' are contested." It is a space, inhabited "not only by those who have migrated and their descendants, but equally by those who are constructed and represented as indigenous" (205).
4. I am referring to reviews by Jenn Blair, Ali Houissa, and Deborah Donovan.
5. I am relying on Soheir Morsey's article, "Beyond the Honorary White Classification for Egyptians: Societal Identity in an Historical Context" (1994).
6. I use the term in an ethnographic sense, where a simple incident "can widen out into enormous complexities of social experience" as suggested by Clifford Geertz in *The Interpretations of Cultures: Selected Essays* (19, 1973).
7. An Egyptian dish made of a mix of rice, pasta, and lentils, topped with tomato hot sauce, and zesty sauce, and garnished with crispy fried onions.
8. Suad Joseph in "Against the Grain of the Nation—the Arab-" (1999) holds that an Arab American is perceived as "not quite American" due to a variety of discourses—specially mainstream media—that present the Arab "as essentially different from the Western, the American" on the bases of the perception of the Arab "as a not-independent, not-autonomous, not individual, not-free person" (257–258).
9. It is noteworthy that despite the detailed descriptions of the buildings in Central Square, Lahiri does not give any description

for any activities, events, or incidents taking place in the neighborhood, despite the fact that during the 1960s (the same time period that Ashima and Ashoke moved to the United States) Central Square witnessed several protests against the war in Vietnam and after the assassination of Martin Luther King in 1968. For the history of Central Square, I am relying on Sarah Boyer's study, *Crossroads: Stories of Central Square Cambridge, Massachusetts 1912–2000* (2001). Lavina Dhingra and Floyd Cheung in *Naming Jhumpa Lahiri: Canons and Controversies* (2011) term this absence of political edge in Lahiri's works "deracination" of characters and "depoliticizing of the historical events" (xvii).

10. I draw on Rosemary George's argument in *The Politics of Home: Postcolonial Relocations and Twentieth-century Fiction* (1996). She holds that in migrant narratives that foreground family history, stories of family history do not "replicate a 'national culture' [but] compensate for the lack of other filiations" in the host country. These stories are neither "a microcosmic or allegorical version of the nation" (190).

4 Transnational Allegories and the Non-national Subject in *The Agüero Sisters* and *The Night Counter*

1. I am referring here to studies by Isabel Alvarez-Borland's *Cuban-American Literature in Exile* (1998); María Cristina García *Havana USA: Cuban Exiles and Cuban Americans in South Florida* (1996); and Teresa Derrickson's *Politicizing Globalization: Transnational Conflict and Change in the Contemporary Novel* (2002).
2. Marti wrote "Flores del Destierro" between 1882 and 1891 and was published posthumously in 1933.
3. Postcolonial critical studies of magic realism can be exemplified by Frederic Jameson's *The Political Unconscious*, Kumkum Sangari's *Politics of the Possible*, and Stephen Slemon's "Magic Realism as Postcolonial Discourse," to name only a few.
4. Labid Ibn Rabia is a pre-Islamic poet known for his odes in the Jahilya era. The Mu'allaqat are very long poems that address various themes; they always start with lamenting deserted dwellings, followed by a description of their ruins, then to courtship, then a description of the desert (and tribal battles), and ends with the poet valorizing his kinfolk.

5. For example, the story where Ibn Rabia's verses first appear in *The Arabian Nights* is the story of *The Third Dervish*, told on the fifty-sixth night, which stands for the inevitability of facing one's destiny (from the 1990 Norton edition translated by Husain Haddawy).
6. Saree Makdisi and Felicity Nussbaum in *The Arabian Nights in Historical Context* (2008) explain that the first translation for the *Nights* was by Antoine Galland in the early eighteenth century and by 1800 there were numerous English editions of its tales (2–3).

Bibliography

Abdulhadi, Rabab, Evelyn Alsultany, and Nadine Naber, eds. *Arab and Arab American Feminisms: Gender, Violence, and Belonging*. Syracuse, NY: Syracuse University Press, 2011.

Abu-Jaber, Diana. *The Language of Baklava*. New York: Anchor Books, 2005.

Ahmed, Sara. *The Cultural Politics of Emotion*. London: Routledge, 2004.

Al-Ali, Nadji and Khalid Koser, eds. *New Approaches to Migration: Transnational Communities and the Transformation of Home*. London and New York: Routledge, 2002.

Alawi, Nabil. "Arab American Poets: The Politics of Exclusion and Assimilation." In *Narratives of Resistance: Literature and Ethnicity in the United States and the Caribbean*, edited by Ana Manzanas and Jesus Benito. Universidad de Castilla La Mancha, 1999. 43–56.

Alvarez-Borland, Isabel. *Cuban-American Literature in Exile: From Person to Persona*. Charlottesville: University Press of Virginia, 1998.

Althusser, Louis. "Ideology and Ideological State Apparatuses (Notes towards an Investigation)." *Lenin and Philosophy and Other Essays*, translated by Ben Brewster. New York and London: Monthly Review Press, 1971. 127–186.

Anderson, Benedict. *Imagined Communities: Reflections on the Origin and Spread of Nationalism*. London and New York: Verso, 1991.

Appadurai, Arjun. "Disjuncture and Difference in the Global Cultural Economy." In *Modernity at Large*. Minneapolis and London: University of Minnesota Press, 1996. 27–47.

Appadurai, Arjun, ed. *The Social Life of Things: Commodities in Cultural Perspectives*. Cambridge and New York: Cambridge University Press, 1986.

Appelbaum, Richard P. "Multiculturalism and Flexibility: Some New Directions in Global Capitalism." In *Mapping Multiculturalism*, edited by Avery F. Gordon, and Christopher Newfield. Minneapolis: University of Minnesota Press, 1996. 297–316.

Arieli, Yehoshua. "Individualism and National Identity." In *American Chameleon: Individualism in Trans-National Context*, edited by

Richard Curry and Lawrence Goodheart. Kent: Kent State University Press, 1991. 167–187.

Augé, Marc. *Non-Places: Introduction to an Anthropology of Supermodernity.* London: Verso, 1995.

Bal, Mieke. "Food, Form, and Visibility: Glub and the Aesthetics of Everyday Life." *Postcolonial Studies* 8, no. 1 (2005): 51–73.

Bangs, Pat. Review of *The Language of Baklava. School Library Journal* 51, no. 9 (September 2005): 246.

Bardenstein, Carol. "Transmissions Interrupted: Reconfiguring Food, Memory, and Gender in the Cookbook-Memoirs of Middle Eastern Exiles." *Signs* 28, no. 1 (2002): 353–387.

———. "Beyond Univocal Baklava: Deconstructing Food-as-Ethnicity and the Ideology of Homeland in Diana Abu-Jaber's *The Language of Baklava.*" *Journal of Arabic Literature* 41 (2010): 160–179.

Barthes, Roland. "Toward a Psychology of Contemporary Food Consumption." In *Food and Culture: A Reader*, edited by Carole Counihan and Penny Van Esterik. New York and London: Routledge, 1997. 20–27.

Behdad, Ali. "Nationalism and Immigration to the United States." *Diaspora* 6, no. 2 (1997): 155–178.

Benjamin, Walter. *The Origin of German Tragic Drama.* London and New York: Verso, 1985.

Bent, Eliza. "Keeping the Faith." *American Theatre* 24, no. 3 (March 2007): 24.

Bhabha, Homi K. *The Location of Culture.* London and New York: Routledge, 1994.

———. "DissemiNation: Time, Narrative, and the Margins of the Modern Nation." In *Nation and Narration*, edited by Homi Bhabha. London: Routledge, 1990. 291–322.

Bhalla, Tamara Ayesha. "Between History and Identity: Reading the Authentic in South Asian Diasporic Literature and Community." Order No. 3343010 University of Michigan, 2008. Ann Arbor: *ProQuest.* Web. 17 Mar. 2015.

Bharucha, Nilufer. "Reflections in Broken Mirrors: Diverse Diasporas in Recent Parsi Fiction." *Wasafiri.* 10, no. 21 (1995a): 32–35.

———. "'When Old Tracks are Lost': Rohinton Mistry's Fiction as Diasporic Discourse." *The Journal of Commonwealth Literature* 30, no. 2 (1995b): 57–63.

Blair, Jenn. Review of *Dinarzad's Children.* MELUS 31, no. 4 (Winter 2006): 272–274.

Blake, Debra. Review of *The Last Generation. The Journal of the Midwest Modern Language Association*, 28, no. 1, Identities (Spring 1995): 99–102.

Bloch, Ernst. *The Heritages of Our Time*. Berkeley: University of California Press, 1991.

Bourne, Randolph. "Trans-National America." In *War and the Intellectuals: Essays by Randolph Bourne 1915–1919*, edited by Carl Resek. New York: Harper and Row, 1964. 107–123.

Boyer, Sarah, ed. *Crossroads: Stories of Central Square Cambridge, Massachusetts 1912–2000*. Cambridge, MA: Cambridge Historical Commission, 2001.

Brady, Mary Pat. *Extinct Lands, Temporal Geographies: Chicana Literature and the Urgency of Space*. Durham, NC and London: Duke University Press, 2002.

Brah, Avtar. *Cartographies of Diaspora: Contested Identities*. New York and London: Routledge, 1996.

Brandeis, Louis. "To 'Possess the National Consciousness of an American.'" *Center for Immigration Studies*. www.cis.org. March 8, 2011.

Brettell, Caroline and James Hollifield, eds. "Migration Theory: Talking across Disciplines." In *Migration Theory: Talking Across Disciplines*. New York: Routledge, 2000. 1–26.

Brodsky, Anne. "Violence against Afghan Women: Tradition, Religion, Conflict, and War." In *Gender and Violence in the Middle East*, edited by Moha Ennaji and Fatima Sadiqi. New York: Routledge, 2011. 115–138.

Bryce, James. *The American Commonwealth*. 2nd ed. vol. 2. London: Macmillan, 1891.

Burnett, Christina Duffy and Burke Marshall eds. *Foreign in a Domestic Sense: Puerto Rico, American Expansion, and the Constitution*. Durham, NC and London: Duke University Press, 2001.

Casanova, Pascale. "Principles of a World History of Literature." *The World Republic of Letters*. Casanova, translated by M. B. DeBevoise. Cambridge: Harvard University Press, 2004. 9–44.

Castronovo, Russ. "Nation dot com: American Studies and the Production of the Corporatist Citizen." In *The Futures of American Studies*, edited by Donald Pease and Robyn Wiegman. Durham, NC and London: Duke University Press, 2002. 536–56.

Certeau, Michel De. *The Practice of Everyday Life*. Berkley and LA: University of California Press, 1984.

Chadwell, Fay A. Review of *West of the Jordan*. Library Journal (April 2003): 128–9.

Cortina, Rodolfo J. "History and Development of Cuban American Literature: A Survey." In *Handbook of Hispanic Cultures in the United States: Literature and Art*, edited by Francisco Lomelí. Houston: Arte Público Press, 1993. 40–61.

Curry, Richard and Lawrence Goodheart, eds. *American Chameleon: Individualism in Trans-National Context*. Kent: Kent State University Press, 1991.

Derrickson, Teresa. *Politicizing Globalization: Transnational Conflict and Change in the Contemporary Novel*. Indiana University of Pennsylvania. *ProQuest Dissertations and Theses*, 274–274. Retrieved from https://ezproxy.lib.uwm.edu/ http: //search. proquest.com /docview/276208191?accountid=15078 (276208191). PhD Diss. Ann Arbor: Indiana University of Pennsylvania, 2002.

DeSpain, Dori. Review of *West of the Jordan*. *School Library Journal* (October 2003): 208.

Dhawan, Rajinder and Novy Kapadia. *The Novels of Bapsi Sidhwa*. New York: Prestige Books, 1996.

Dhingra, Lavina and Floyd Cheung. *Naming Jhumpa Lahiri: Canons and Controversies*. Maryland: Lexington, 2011.

Dirlik, Arif. "The Past as Legacy and Project: Postcolonial Criticism in the Perspective of Indigenous Historicism." *American Indian Culture and Research Journal* 20 (1996): 1–31.

Donovan, Deborah. Review of *Dinarzad's Children*. *Booklist* 101, no. 6 (2004): 560–561.

Duany, Jorge. "Neither Golden Exile nor Dirty Worm: Ethnic Identity in Recent Cuban-American Novel." *Cuban Studies* 23 (1993): 167–83.

Elia, Nada. "Islamophobia and the 'Privileging' of Arab American Women." *NWSA Journal* 18, no. 3 (Fall 2006): 155–61.

Engberg, Gillian. Review of *The Language of Baklava*. *Booklist* 101, no. 12 (2005): 1051.

Fadda-Conrey, Carol. "Arab American Literature in the Ethnic Borderland: Cultural Intersections in Diana Abu-Jaber's *Crescent*." *MELUS* 31, no. 4 (Winter 2006): 187–205.

Field, Robin. "A Prophet in Her Own Town: an Interview with Diana Abu-Jaber." *MELUS* 31, no. 4 (Winter 2006): 207–225.

Franco, Jean. "The Nation as Imagined Community." In *Dangerous Liaisons: Gender, Nation, and Postcolonial Perspectives*, edited by Anne McClintock, Aamir Mufti, and Ella Shohat, eds. Minneapolis: University of Minnesota Press, 1997, 130–137.

Friedman, Susan. "Unthinking Manifest Destiny: Muslim Modernities on Three Continents." In *Shades of the Planet: American Literature as World Literature*, edited by Wai Dimock and Lawrence Buell. Princeton: Princeton University Press, 2007, 62–99.

Fusco, Coco. *English Is Broken Here: Notes on Cultural Fusion in the Americas*. New York: The New Press, 1995.

Gabaccia, Donna. *From the Other Side: Women, Gender, and Immigrant Life in the U.S., 1820–1990.* Minneapolis: Indiana University Press, 1994.

———. *We Are What We Eat: Ethnic Food and the Making of Americans.* Cambridge: Harvard University Press, 1998.

Garcia, Cristina. *The Agüero Sisters.* New York: Alfred Knopf, 1997.

García, María Cristina. *Havana USA: Cuban Exiles and Cuban Americans in South Florida, 1959–1994.* Berkley and Los Angeles: University of California Press, 1996.

Gaztambide, Elsa. Review of *West of the Jordan. Booklist* (June 1–15, 2003).

Geertz, Clifford. *The Interpretations of Cultures: Selected Essays.* New York: Basic Books, 1973.

George, Rosemary Marangoly. *The Politics of Home: Postcolonial Relocations and Twentieth-Century Fiction.* Berkley, Los Angeles, and London: University of California Press, 1996.

———. "But That Was in Another Country: Girlhood and the Contemporary 'Coming to America' Narrative." In *The Girl: Constructions of the Girl in Contemporary Fiction by Women*, edited by Ruth Saxton. New York: St. Martin Press, 1998. 135–152.

Giddens, Anthony. *The Consequences of Modernity.* Stanford University Press, 1990.

Grewal, Inderpal and Caren Kaplan, eds. *Scattered Hegemonies: Postmodernity and Transnational Feminist Practices.* 4th ed. Minneapolis: University of Minnesota Press, 2002.

Guzmán, Isabel Molina. "Disorderly Bodies and Discourses of Latinidad in the Elián González Story." In *From Bananas to Buttocks: The Latina Body in Popular Film and Culture*, edited by Myra Mendible. Austin: University of Texas Press, 2007. 219–42.

Habermas, Jürgen. *The Structural Transformation of the Public Sphere: an Inquiry into a Category of Bourgeois Society.* translated by Thomas Burger. Cambridge: MIT Press, 1989.

Hai, Ambreen. "Border Work, Border trouble: Postcolonial Feminism and the Ayah in Bapsi Sidhwa's *Cracking India.*" *Modern Fiction Studies.* 46, no. 2 (2000): 379–426.

Halaby, Laila. *West of the Jordan.* Boston, MA: Beacon Press, 2003.

———. An Online Interview with the author. http://www.beacon.org/Assets/ClientPages/WestOfTheJordanrg.aspx http://www.lailaHalaby.net/guides/

Hall, Stuart. "Encoding/decoding." In *Culture, Media, Language*, edited by Stuart Hall, Dorothy Hobson, Andrew Lowe, and Paul Willis, eds London: Routledge, 2005. 107–116.

Haney-Lopez, Ian. *White by Law: The Legal Construction of Race*. New York and London: New York University Press, 2006.

Harootunian, Harry. "Some Thoughts on Comparability and the Space-Time Problem." *Boundary* 232, no. 2 (2005): 23–52.

Harris, Joy. Review of *The Language of Baklava. Publishers Weekly* 252, no. 3 (2005): 43.

Hassan, Wael. *Immigrant Narrative: Orientalism and Cultural Translation in Arab American and Arab British Literature*. New York: Oxford University Press, 2011.

Hirsch, Marianne. *The Mother/Daughter Plot: Narrative, Psychoanalysis, Feminism*. Bloomington and Indianapolis: Indiana University Press, 1989.

Hobsbawm, E.J. *Nations and Nationalism Since 1780: Programme, Myth, Reality*. Cambridge: Cambridge University Press, 1990.

Hong, Grace Kyungwon and Roderick Ferguson, eds. *Strange Affinities: The Gender and Sexual Politics of Comparative Racialization*. Durham, NC and London: Duke University Press, 2011.

Honig, Bonnie. *Democracy and the Foreigner*. Princeton and Oxford: Princeton University Press, 2001.

Hooglund, Eric. *Crossing the Waters*. Smithsonian Institution Press, 1987.

Houissa, Ali. Review of *Dinarzad's Children. Library Journal*. 129, no. 20 (2004): 106.

Ian, Buchanan. "National Allegory Today: A Return to Jameson." In *On Jameson: From Postmodernism to Globalism*, edited by Caren Irr and Ian Buchanan. Albany: State University of New York Press, 2006. 173–188.

Imre, Szeman. "Who's Afraid of National Allegory? Jameson, Literary Criticism, Globalization." In *On Jameson: From Postmodernism to Globalism*, edited by Caren Irr and Ian Buchanan. Albany: State University of New York Press, 2006. 189–211.

Irr, Caren and Ian Buchanan, eds. *On Jameson: From Postmodernism to Globalism*. Albany: State University of New York Press, 2006.

Jacobson, Mathew Frye. *Whiteness of a Different Color: European Immigrants and the Alchemy of Race*. Cambridge: Harvard University Press, 1998.

Jameson, Fredric. "Third World Literature in the Era of Multinational Capitalism." *Social Text*, 15 (Autumn 1986): 65–88.

———. *The Cultural Turn: Selected Writings on the Postmodern 1983–1998*. London: Verso, 1998.

———. *The Political Unconscious: Narrative as a Socially Symbolic Act*. London: Routledge, 2002.

Jameson, Fredric and Buchanan Ian, eds. *Jameson on Jameson: Conversations on Cultural Marxism*. Durham, NC and London: Duke University Press, 2007.

Japtok, Martin. *Growing Up Ethnic: Nationalism and the Bildungsroman in African American and Jewish American Fiction*. Iowa: University of Iowa Press, 2005.

Jay, Paul. "Beyond Discipline? Globalization and the Future of English." PMLA, Vol. 116, 1, Special Topic: Globalizing Literary Studies (January 2001): 32–47.

Joseph, Suad. "Against the Grain of the Nation—The Arab-." In *Arabs in America: Building a New Future*, edited by Michael Suleiman. Philadelphia: Temple University Press, 1999. 257–271.

Kadi, Joanna, ed. *Food for Our Grandmothers: Writings by Arab-American and Arab-Canadian Feminists*. Boston, MA: South End Press, 1994.

Kahf, Mohja. "Packaging 'Huda': Sha'rawi's Memoirs in the United States Reception Environment." In *Going Global: The Transnational Reception of Third World Women Writers*, edited by Amal Amireh and Lisa Suhair Majaj. New York and London: Garland Publishing, 2000. 148–172.

Kaldas, Pauline. *The Time between Places: Stories that Weave in and out of Egypt and America*. Fayetteville: University Arkansas Press, 2010.

———. Personal Interview. January 2013.

Kandiyoti, Dalia. *Migrant Sites: America, Place and Diaspora Literatures*. New Hampshire: Dartmouth College Press, 2009.

Kaplan, Amy. "Manifest Domesticity." In *The Anarchy of Empire: in the Making of U.S. Culture*. Cambridge: Harvard University Press, 2002. 23–52.

———. "Violent Belongings and the Question of Empire Today Presidential Address to the American Studies Association, Hartford, Connecticut, October 17, 2003." *American Quarterly* 56, no. 1 (March 2004): 1–18.

King, Adele. Review of *An American Brat. World Literature Today*. Indian Literatures: In the Fifth Decade of Independence. 68, no. 2 (Spring 1994): 436.

Kristeva, Julia. "Women's Time." In *The Kristeva Reader*, edited by Toril Moi. New York: Columbia University Press, 1986. 187–213.

Kysar, Kathryn. Review of *The Night Counter. The Minneapolis Star Tribune*. http://www.startribune.com/entertainment/book August 8, 2009.

Lahiri, Himadri. "Individual-Family Interface in Jhumpa Lahiri's *The Namesake*." Americana. IV, 2(Fall 2008) e-journal http://americanaejournal.hu/vol4no2/lahiri

Lahiri, Jhumpa. *The Namesake*. Boston, MA: Houghton Mifflin, 2003.

Leong, Karen. *The China Mystique: Pearl S. Buck, Anna May Wong, Mayling Soong and the Transformation of American Orientalism*. Berkeley: University of California Press, 2005.

Levander, Caroline Field and Robert S. Levine, eds. *Hemispheric American Studies*. New Brunswick, NJ: Rutgers University Press, 2008.

Levitt, Peggy. "Social Remittances: Migration Driven Local-Level Forms of Cultural Diffusion." *International Migration Review* 32, no. 4 (Winter 1998): 926–948.

Levitt, Peggy and Nina Schiller. "Conceptualizing Simultaneity: A Transnational Social Field Perspective on Society." In *The Transnational Studies Reader: Intersections and Innovations*, edited by Sanjeev Khgram and Peggy Levitt. New York and London: Routledge, 2008.

Lowe, Lisa. *Immigrant Acts: On Asian American Cultural Politics*. Durham, NC and London: Duke University Press, 1999.

Ludescher, Tanyss. "From Nostalgia to Critique: An Overview of Arab American Literature." *MELUS* 31, no. 4 (Winter 2006): 93–114.

Lutwack, Leonard. *The Role of Place in Literature*. Syracuse, NY: Syracuse University Press, 1994.

Majaj, Lisa Suhair. "Arab-Americans and the Meaning of Race." In *Postcolonial Theory and the United States: Race, Ethnicity, and Literature*, edited by Amritjit Singh and Peter Schmidt. Jackson: University Press of Mississippi, 2000. 320–337.

Makdisi, Saree and Felicity Nussbaum, eds. *The Arabian Nights in Historical Context: Between East and West*. Oxford and New York: Oxford University Press, 2008.

Mamdani, Mahmood. *Good Muslim, Bad Muslim: America the Cold War, and the Roots of Terror*. New York: Three Leaves Press, 2004.

Mannur, Anita. "Culinary Nostalgia: Authenticity, Nationalism, and Diaspora." *MELUS* 32, no. 4 (Winter 2007): 11–31.

Massey, Doreen. *Space, Place, and Gender*. Minneapolis: University of Minnesota Press, 1994.

McCook, Stuart George. *States of Nature: Science, Agriculture, and Environment in The Spanish Caribbean 1760–1940*. Austin: University of Texas Press, 2002.

McLaren, Peter. "White Terror and Oppositional Agency: Towards a Critical Multiculturalism." In *Multiculturalism: A Critical Reader*, edited by David Theo Goldberg. Oxford: Blackwell, 1994. 45–74.

McNulty, Lori. *Contagious Poetics: Rumor, Ritual and Resistance in Zora Neale Hurston's "Tell My Horse." ProQuest Dissertations and*

Theses, 128–128. Retrieved from https://ezproxy.lib.uwm.edu/login?url=http://search.proquest.Com/docview/304560599?accountid=15078. (304560599).

Meyerhoff, Hans. *Time in Literature*. Berkeley and Los Angeles: University of California Press, 1955.

Miles, Robert. *Racism and Migrant Labour*. London and Boston, MA: Routledge, 1982.

Mishra, Vijay. *Literature of the Indian Diaspora: Theorizing the Diasporic Imaginary*. New York: Routledge, 2006.

Morace, Robert A. Review of *An American Brat. Magill Book Reviews*. 10/01/1994.

Moraga, Cherríe. *The Last Generation*. Boston, MA: South End Press, 1993.

Morsey, Soheir. "Beyond the Honorary White Classification for Egyptians: Societal Identity in an Historical Context." In *Race*, edited by Steve Gregory and Roger Sanjek. New Brunswick: Rutgers University Press, 1994. 175–198.

Murji, Karim and John Solomos, eds. *Racialization: Studies in Theory and Practice*. New York: Oxford University Press, 2005.

Naber, Nadine. "Ambiguous Insiders, an Investigation of Arab American Invisibility." *Journal of Ethnic and Racial Studies*. 23, no. 1 (2000): 37–61.

Naff, Alixa. *Becoming American: The Early Arab Immigrant Experience*. Carbondale: Southern Illinois University, 1993.

———. *The Arab Americans*. Philadelphia, PA: Chelsea House Publications, 1999.

Naficy, Hamid. *The Making of Exile Cultures: Iranian Television in Los Angeles*. Minneapolis: University of Minnesota Press, 1993.

———. *An Accented Cinema: Exilic and Diasporic Filmmaking*. Princeton, NJ: Princeton University Press, 2001.

Nash, Walter. *Language in Popular Fiction*. London and New York: Routledge, 1990.

Nava, Mica. "The Allure of Difference: Selfridges, the Russian Ballet and the Tango." In *Visceral Cosmopolitanism: Gender, Culture and the Normalisation of Difference*. edited by Mica Nava. Oxford and New York: Berg, 2007.

Ní Laoire, Caitríona. "Return Migrants and Boundaries of Belonging." In *Globalization, Migration, and Social Transformation: Ireland in Europe and the World*, edited by Bryan Fanning and Ronaldo Munck. Burlington: Ashgate, 2011. 21–34.

Onega, Susana. *Jeanette Winterson*. Manchester and New York: Manchester University Press, 2006.

Oweiss, Ibrahim. "Petrodollar Surplus: Trends and Economic Impact." *Journal of Energy and Development.* 9, no. 2 (1984). 177–203.

Owens, Craig. "The Allegorical Impulse: Toward a Theory of Postmodernism." In *Art in Theory 1900–1990: An Anthology of Changing Ideas*, edited by Charles Harrison and Paul Wood. Oxford: Blackwell, 1992. 1051–1060.

Page, Kezia. *Transnational Negotiations in Caribbean Diasporic Literature.* New York and London: Routledge, 2011.

Pease, Donald. "National Identities, Postmodern Artifacts, and Postnational Narratives." Boundary 2, 19, 1 *New Americanists 2:* National Identities and Postnational Narratives (Spring 1992): 1–13.

———. "C.L.R. James, Moby-Dick and the Emergence of Transnational American Studies." In *The Futures of American Studies*, edited by Donald Pease and Robyn Wiegman. Durham, NC: Duke University Press, 2002. 135–163.

Pease, Donald E. and Amy Kaplan, eds. *Cultures of United States Imperialism.* Durham, NC and London: Duke University Press, 1993.

Pease, Donald E. and Robyn Wiegman. *The Futures of American Studies.* Durham, NC and London: Duke University Press, 2002.

Penenberg, Adam. "Immigrant Encounters an American Brat." *New York Times Book Review* 1, no. 16 (1994).

Pérez-Torres, Rafael. *Movements in Chicano Poetry: Against Myths, Against Margins.* New York: Cambridge University Press, 1995.

———. "Refiguring Aztlán." In *Postcolonial Theory and the United States: Race, Ethnicity, and Literature*, edited by Amritjit Singh and Peter Schmidt. Jackson: University Press of Mississippi, 2000. 103–121.

Prashad, Vijay. *The karma of Brown Folk.* Minneapolis: University of Minnesota Press, 2000.

———. "Orientalism." In *Keywords for American Cultural Studies*, edited by Bruce Burgett and Glenn Hendler. New York: University of New York Press, 2007. 174–177.

Puar, Jasbir. *Terrorist Assemblages: Homonationalism in Queer Times.* Durham, NC and London: Duke University Press, 2007.

Radhakrishnan, R. "Is Ethnic 'Authentic' in the Diaspora?" In *The State of Asian America: Activism and Resistance in the 1990s*, edited by Karin Aguilar-San Juan. Cambridge: South End Press, 1994. 219–234.

Radway, Janice. *Reading the Romance: Women, Patriarchy, and Popular Literature.* Chapel Hill and London: University of North Carolina Press, 1984.

———. "What's in a Name." In *The Futures of American Studies*, edited by Donald Pease and Robyn Wiegman. Durham, NC and London: Duke University Press, 2002. 45–75.

Reddy, Chandan. *The Migrant Present: Alienage, Race and the Politics of Black Internationalism, 1898–1953*. Columbia University. *ProQuest Dissertations and Theses*, 327–327. Retrieved from https://ezproxy.lib.uwm.edu/login?url=http://search.proquest.com/ docview/305207654?accountid=15078. (305207654).

Ricoeur, Paul. *Time and Narrative*. Vol. 2 translated by Kathleen McLaughlin and David Pellauer. Chicago and London: University of Chicago Press, 1984.

Rowe, John Carlos. "Postnationalism, Globalism, and the New American Studies." In *The Futures of American Studies*, edited by Donald Pease and Robyn Wiegman. Durham, NC and London: Duke University Press, 2002. 167–182.

Rutherford, Jonathan. *Identity: Community, Culture, Difference*. London: Lawrence and Wishart, 1990.

Said, Edward W. *Culture and Imperialism*. New York: Knopf, 1993.

———. *Reflections on Exile*. Cambridge: Harvard University Press, 2001.

Salaita, Steven. *Arab American Literary Fictions, Cultures, and Politics*. New York: Palgrave, 2007.

———. *Modern Arab American Fiction: A Reader's Guide*. New York: Syracuse University Press, 2011.

Sangari, Kumkum. "Politics of the Possible: or the Perils of Reclassification." *Politics of the Possible: Essays on Gender, History, Narrative, Colonial English*. New Delhi: Tulika, 2001. 1–28.

———. "Ruptures, Junctures, Returns: Unveiled History." In *Trace, Retrace*, edited by Kumkum Sangari, New Delhi: Tulika, 2013. 114–183.

Sarkar, Parama. Review of *West of the Jordan*. MELUS 31, no. 4 (Winter 2006): 263–265.

Sennett, Richard. "The Public Realm." *The Blackwell City Reader*, edited by Gary Bridge and Sophie Watson. Massachusetts: Blackwell Publishing Ltd, 2010. 261–272.

Shakir, Evelyn. "Arab-American Literature." In *New Immigrant Literatures in the United States: A Source-book to Our Multicultural Literary Heritage*, edited by Alpana Knippling. Westport, CT: Greenwood Press, 1996. 3–18.

———. *Bint Arab: Arab and Arab American Women in the United States*. Westport, CT: Praeger, 1997.

Shankar, Lavina Dhingra. "The Limits of (South Asian) Names and Labels: Postcolonial or Asian American?" In *A Part, Yet Apart: South Asians in Asian America*, edited by Shankar and Srikanth. Philadelphia: Temple University Press, 1998. 49–66.

Shankar, Lavina Dhingra. "Not Too Spicy: Exotic Mistresses of Cultural Translation in the Fiction of Chitra Divakaruni and Jhumpa Lahiri." In *Other Tongues: Rethinking the Language Debates in India*, edited by Nalini Iyer and Bonnie Zare. Amsterdam and New York: Rodopi, 2009. 23–52.

Shankar, Lavina Dhingra and Rajini Srikanth, eds. *A Part, Yet Apart: South Asians in Asian America.* Philadelphia: Temple University Press, 1998.

Shankar, Rajiv. "South Asian Identity in Asian America." Foreword. In *A Part, Yet Apart: South Asians in Asian America*, edited by Shankar and Srikanth. Philadelphia: Temple University Press, 1998. ix–xv.

Sidhwa, Bapsi. *An American Brat.* Minnesota: Milkweed, 2006.

Sidhwa, Bapsi and Preeti Singh. "My Place in the World." *Alif: Journal of Comparative Poetics.* Post-Colonial Discourse in South Asia 18 (1998): 290–298.

Sihon, Winifred. Review of *An American Brat.* Book Report 13, no. 5 (March/April 95): 41.

Simmel, Georg. "The Stranger." In *Sociology of Georg Simmel*, edited by and translated by Kurt Wolff. Glencoe: The Free Press, 1950. 402–408.

Singh, Ranghi. *Bapsi Sidhwa.* New Delhi: IVY Publishing House, 2005.

Slemon, Stephen. "Magic Realism as Postcolonial Discourse." In *Magic Realism: Theory, History, Community*, edited by Lois Parkinson Zamora and Wend B. Faris. Durham, NC and London: Duke University Press, 1995. 407–426.

Smith, Anthony. *Myths and Memories of the Nation.* Oxford and New York: Oxford University Press, 1999.

Soto, Sandra. "Where in the Transnational World are U.S. Women of Color." In *Women Studies for the Future: Foundations, Interrogations, Politics*, edited by Elizabeth Lapovesky Kennedy and Agatha Beins. New Jersey: Rutgers University Press, 2005. 111–124.

———. "Transnational Knowledge Projects and Failing Racial Etiquette." National Association for Chicana and Chicano Studies Annual Conference, 2008. 69–72.

Spivak, Gayatri. "Three Women's Texts and a Critique of Imperialism." In *'Race,' Writing, and Difference*, edited by Henry Louis Gates Jr. Chicago and London: Chicago University Press, 1985. 262–280.

Stanley, Sandra Kumamoto, ed. *Other Sisterhoods: Literary Theory and U.S. Women of Color.* Urbana: University of Illinois Press, 1998.

Stephanson, Anders. *Manifest Destiny: American Expansionism and the Empire of Right.* New York: Hill and Wang, 1995.

———. “Homecomings to the Future: From Diasporic Mythographies to Social Projects of Return.” In *Homecomings: Unsettling Paths of Return*, edited by Fran Markowitz and Anders Stefansson. Maryland: Lexington, 2004. 2–20.

Strehle, Susan. *Transnational Women's Fiction: Unsettling Home and Homeland*. New York: Palgrave, 2008.

Sugg, Katherine. *Gender and Allegory in Transamerican Fiction and Performance*. New York: Palgrave, 2008.

Suleiman, Michael, ed. *Arabs in America: Building a New Future*. Philadelphia: Temple University Press, 1999.

Takaki, Ronald. “Reflections on Racial Patterns in America.” In *From Different Shores: Perspectives on Race and Ethnicity in America*, edited by Ronald Takaki. New York and Oxford: Oxford University Press, 1987. 26–37.

———. *Strangers from a Different Shore: A History of Asian Americans*. Boston, MA: Little Brown, 1989.

Tatonetti, Lisa. “‘A Kind of Queer Balance’: Cherríe Moraga's Aztlán.” MELUS, 29, no. 2 (Summer 2004): 227–247.

Thomas, Devon. “*The Language of Baklava*.” *Library Journal* 130, no. 2 (2005): 109.

Tompkins, Kayla Wazana. “History's Traces: Personal Narrative, Diaspora, and the Arab Jewish Experience.” In *Arab and Arab American Feminisms: Gender, Violence, and Belonging*, edited by Rabab Abdulhadi, Evelyn Alsultany, and Nadine Naber. Syracuse, NY: Syracuse University Press, 2011. 124–137.

Uchida, Aki. “The Orientalization of Asian Women in America.” *Women Studies International Forum* 21, no. 2 (1998): 161–174.

Urry, John. *The Tourist Gaze*. 2nd ed. London: Sage Publications, 2002.

Varas, Patricia Eugenia. *The New Narrative and the National Popular Culture in Ecuador: Three Representative Narrators*. University of Toronto (Canada). *ProQuest Dissertations and Theses*, 263–263. Retrieved from https://ezproxy.lib.uwm.edu/ login?url=http:// search. proquest.com/docview/303910115?accountid=15078. (303910115).

Volpp, Leti. “The Citizen and the Terrorist.” In *American Studies: An Anthology*, edited by Janice Radway, Kevin Gaines, Barry Shank and Penny Von Eschen. Chichester, U.K.; Malden, MA New York: Wiley-Blackwell, 2009. 78–88.

Webster, Yehudi. *The Racialization of America*. New York: St. Martin's Press, 1992.

Wiegman, Robyn. *Object Lessons.* Durham, NC and London: Duke University Press, 2012.

Williams, Raymond. *Keywords: A Vocabulary of Culture and Society.* New York: Oxford University Press, 1983.

Wilson, Rob. "Techno-euphoria and the Discourse of the American Sublime." In *National Identities and Post-Americanist Narratives*, edited by Donald E. Pease. Durham, NC and London: Duke University Press, 1994. 205–229.

Wilson, Rob and Wimal Dissanayake, eds. *Global/Local: Cultural Production and the Transnational Imaginary.* Durham, NC: Duke University Press, 1996.

Wilson, Woodrow. "Address to Naturalized Citizens at Conventional Hall, Philadelphia, May 10, 1915." *The American Presidency Project.* www.presidency.ucsb.edu. March 21, 2011.

Winterson, Jeanette. *Oranges Are Not the Only Fruit.* Introduction. London: Vintage, 1991.

Yunis, Alia. *The Night Counter.* New York: Shaye Areheart Books, 2009.

Zizek, Slavoj. "Multiculturalism, or, the Cultural Logic of Multinational Capitalism." *New Left Review* Issue 225 (1997): 28–51.

Websites

www.aaiusa.org/arab-americans

www.quickfacts.census.gov/qfd/meta/long_68178

Index

CPSIA information can be obtained at www.ICGtesting.com
Printed in the USA
BVOW06*0213270416

445772BV00004B/12/P

9 781137 502865